SpringerBriefs in Business

More information about this series at http://www.springer.com/series/8860

Veljko Milutinovic · Jakob Salom

Mind Genomics

A Guide to Data-Driven Marketing Strategy

Springer

Veljko Milutinovic
Scientific Advisory Board
Maxeler Technologies
London
United Kingdom

Jakob Salom
Mathematical Institute
Serbian Academy of Sciences and Arts
Belgrade
Serbia

ISSN 2191-5482 ISSN 2191-5490 (electronic)
SpringerBriefs in Business
ISBN 978-3-319-39731-3 ISBN 978-3-319-39733-7 (eBook)
DOI 10.1007/978-3-319-39733-7

Library of Congress Control Number: 2016941600

Printed on acid-free paper

This Springer imprint is published by Springer Nature
The registered company is Springer International Publishing AG Switzerland

Foreword by Howard Moskowitz

Mind Genomics is a new, data-driven approach to understand the world of the everyday. The organizing principle is that knowledge is developed in a pointillist style, from the intimate, profound, and comprehensive understanding of specific topics, be these situations (e.g., eating a breakfast), topics (e.g., ethics and practice of digital piracy), products (e.g., a yogurt), and so forth.

Mind Genomics is best conceived as a way to discover and then to deeply understand what aspects of the topic are important to people, how people differ in the way they respond to these aspects of the topic, and how one goes about identifying these different viewpoints for specific topics.

What this means is that it is now possible to understand people in a deep way and communicate with people in the language that they understand, in a language which motivates them. Whether we are dealing with education (how students want to learn), or medicine (how patients need to interact with the medical community), or commerce (how to advertise to customers), we can base what we do on solid, reproducible, publishable, and archival science. In other words, we can base it on the standard scientific principles that professionals around the world accept.

Media power is shifting from the traditional to the digital landscape. Communication such as the increased use of social media and search engines to target particular groups of people becomes more important. Content personalization is another important trend. Companies need to create personalized content tailored to a customer's unique interests to capture attention. Programmatic advertising has become much more targeted messaging as marketers have the ability to pinpoint specific consumer attributes.

These trends are led by technology development and changing consumer behavior. Today, marketers spend more time on pulling in customers, not pushing customers. Customers do not want overwhelming information. They just want the information they need.

All of these trends need to address the right message to targeted group of people. Mind Genomics can help to identify what to say, how to say it, and to whom. It will help businesses to improve the fundamental return. Our data suggest that the increases in return can be of the order of 10 % to a high as 100 %. It is not the magic of Mind Genomics as a machine. It is rather that the business, the school, and the medical system all improve their performance when they know exactly what to say to each person, appropriate for the person's interests, and well-being.

Howard Moskowitz

Preface

Mind Genomics was first introduced by Dr. Howard Moskowitz, an alumnus of Harvard University and the father of Horizontal Segmentation—a widely accepted business model for targeted marketing and profit maximization. He is a Sigma Xi Laureate and is portrayed by many as a genius (Google: Malcolm Gladwell about Howard Moskowitz). Later on, Howard teamed up with Mr. Stephen Onufrey, who spent most of his professional life at IBM, with Stephen contributing important additions to the business aspects of Mind Genomics. It is the talent of the two of them, working in synergy, which created the major successes of Mind Genomics, mostly in the USA. It is the hope of the authors of this book to bring the worldwide attention to the promise of Mind Genomics, for business in particular, and society in general. The authors of this book, Dr. Veljko Milutinovic and Mr. Jakob Salom, are thankful to Dr. Howard Moskowitz and Mr. Stephen Onufrey for sharing with them the history and specifics about early developments in Mind Genomics and for generously giving them the opportunity and support to do research in the domains of engineering and scientific aspects of the overall system.

Consequently, this book is divided into two parts, one on the major engineering aspects and the other on the related scientific research. The epilogue of the book gives pointers to the most recent work of Dr. Howard Moskowitz and Mr. Stephen Onufrey, related to issues of interest for business and society.

Contents

About the Authors

Veljko Milutinovic

Life Member of the ACM
Fellow Member of the IEEE
Member of Academia Europaea
Member of the Serbian Academy of Engineering
Member of the Scientific Advisory Board of Mind Genomics
Member of the Scientific Advisory Board of Maxeler Technologies

Veljko Milutinovic received his Ph.D. from the University of Belgrade, spent about a decade on various faculty positions in the USA (mostly at Purdue University), and was a codesigner of the DARPA's first GaAs RISC microprocessor. Later he taught and conducted research at the University of Belgrade at the School of Electrical Engineering, the School of Mathematics, the School of Physical Chemistry, and the School of Business Administration. His research is mostly in DataMining algorithms and DataFlow computing, with the stress on mapping of data analytics algorithms onto fast energy efficient architectures. For 7 of his books, forewords were written by 7 different Nobel Laureates with whom he cooperated on his past industry sponsored projects. He has over 40 IEEE or ACM journal papers, over 400 WoS citations, and over 4000 Google Scholar citations.

Jakob Salom

Consultant at The Mathematical Institute of the Serbian Academy of Sciences
Consultant to the Vienna Congress COMSULT
Consultant at Hypo Alpe-Adria-Bank a.d. Belgrade, Serbia
Consultant at Alfa Bank AD Skopje, Macedonia
Vice President for Strategic Development of IPSI, Belgrade, Serbia
President of the Jewish Community of Belgrade, Serbia

Jakob Salom is the President of the Jewish Community of Belgrade, Serbia, consultant at the Mathematical Institute of the Serbian Academy of Sciences, consultant at Hypo Alpe-Adria-Bank a.d. Belgrade, Serbia, consultant at Alfa Bank

AD Skopje, Macedonia, and the Vice President for Strategic Development of IPSI, Belgrade, Serbia. He graduated from the School of Electrical Engineering, University of Belgrade (1975). During his tenure in the banking industry, his teams developed successful statewide payment system solutions, retail banking applications, and a number of sophisticated e-banking solutions. In recent years, he has been working as a consultant in the fields of data warehousing, data archiving, and data mining for banking industry. His innovations were successfully applied in business development processes. He is now involved in banking and insurance industry projects using Mind Genomics and Dataflow Maxeler Technologies. His current research is mostly in business processes for the Internet and the ubiquitous computing.

About Howard Moskowitz

Howard Moskowitz is a well-known experimental psychologist in the field of psychophysics (the study of perception and its relation to physical stimuli) and an inventor of world-class market research technology. Among his important contributions to market research is his 1975 introduction of psychophysical scaling and product optimization for consumer product development. Whereas these methods are standard and well accepted today, they required a massive culture change in the 1975 business community. He is best known for the detailed study he made of the types of spaghetti sauce and horizontal segmentation. By providing a large number of options for consumers, Moskowitz pioneered the idea of intermarket variability as applied to the food industry. In the 1980s, his contributions in sensory analysis were extended to health and beauty aids.

He is the CEO of iNovum Inc. as well as president of Moskowitz Jacobs Inc., a firm he founded in 1981. Dr. Moskowitz graduated from Harvard University in 1969 with a Ph.D. in experimental psychology. Prior to that, he graduated from Queens College (New York), Phi Beta Kappa, with degrees in mathematics and psychology. He has written/edited more than 25 books, has published well over 400 articles, and has lectured in the USA and abroad, and now, he serves on the editorial board of major journals.

His book with co-author Alex Gofman, **Selling Blue Elephants**, demonstrates and popularizes how IdeaMap (iNovum's flagship product) creates new products and messages from areas as diverse as credit cards, jewelry offers, presidential messaging during election years, stock market communications, and transnational innovation.

Dr. Moskowitz has won numerous awards, among them the Scientific Director's Gold Medal for outstanding research at the US Army Natick Laboratories, and the 2001, 2003, and 2004 awards by ESOMAR (European, now World Society Of Market Research) for his innovation in web-enabled, self-authored conjoint measurement and for weak signals research in new trends analysis and concept development. In 2004, Dr. Moskowitz was elected as an IFT Fellow and was also awarded the "David R. Peryam Award," from ASTM, in recognition of outstanding contributions to the field of basic and applied sensory science. In 2005, he was awarded the Charles Coolidge Parlin Marketing Research Award for his substantial contributions and dedication to the advancement of marketing research practices. In 2006, he was the recipient of the ARF Research Innovation Award and The Market Research Council Hall of Fame Award. Moskowitz Jacobs Inc's Mind Genomics(R)/IdeaMap(R).net, received the 2012 Edison Bronze Award for New Product Applied Technology Research Tool. The awards recognize ideas at the forefront of new products, services, marketing, design, and innovation.

Introduction

Books like this one are targeting different audiences: engineers, business people, politicians, etc. We have to look at what motivates each audience. Engineers, academics, and scientists are motivated by intellectual content and explanation. If the process is not masterfully crafted, they find no elegance in the result, even though solid measured results are in place. The engineer/academician/scientist focuses on the design and process, not the results. He/she typically does not worry about the ultimate financial viability of an endeavor. On the other hand, business people and politicians are results oriented. As part of the results, they want to insure that the process is as efficient as possible to keep costs of getting to the end in check in order to produce a cost/benefit analysis in the "black"—profitable. They will want to know what they need to do in order to make it happen, but they do not care about the science or process "under the covers" but do care about the process that their people will have to use in order to bring this to life. They begin with the end in mind.

The conclusion is that there are two separate writing work products which target two separate audiences. One target group consists of people who want to intellectually understand how something works and are not concerned in the financial viability of the idea. The other target group consists of those who want to understand in order to use the idea, and to get someone to embrace their product, service, or idea. This book is aimed for the first target group.

Chapter 1
Introduction to Basic Concepts of Mind Genomics

Abstract This chapter gives a detailed overview of the Mind Genomics process, from the implementation point of view. It defines all relevant terms of interest for phases ONE and TWO of the process, and it marks essential differences between the usual approaches to targeted marketing and the Mind Genomics based approach to targeted marketing. Issues are discussed like: How to prepare the polling campaign that determines the mind types on the market, and how to do all the steps needed to place new prospects (potential customers) into one of the existing mind types.

Keyword Mind Genomics process · Targeted marketing · The polling campaign · Mind types · Linear regression · Classification

Marketing can be broadband or targeted. Broadband marketing means that all customers get the same message, via radio or TV, Internet or newspapers, billboards or taxicabs. Targeted marketing means that each customer receives a different message, tuned to the real needs and mental characteristics of that particular customer.

Targeted marketing is nothing new; it existed for centuries. For example, in medieval times, in a village shop, where the owner knows all the villagers, which is a limited number, to some of them the owner offers guns, and jewelry to the others. What is new, and originated with the appearance of the Internet, is targeted marketing spanning a huge amount of persons unknown to the vendor. This paradigm is referred to, by many, as modern targeted marketing.

Most approaches to modern targeted marketing imply the following drawbacks:

a. At the entry into the system, customers are asked to fill forms.
   This is time consuming and represents an attack on the privacy of individuals.
b. During the exploitation, customers are tracked.

V. Milutinovic and J. Salom, *Mind Genomics*, SpringerBriefs in Business,
DOI 10.1007/978-3-319-39733-7_1

For example, if bacon is purchased in one store, than, when the customer appears at the web portal of another store in the same chain, he/she is again offered bacon, which is annoying.

c. When placing offers to customers, the system is based on socioeconomic principles. Rich customers are offered one set of things, poor the other.

Mind Genomics, which is the main subject of this book, does not have any of the above mentioned drawbacks. It does not ask the customers to fill forms at the entry into the system, it does not track the customers, nor it divides them along the socioeconomic division lines; yet, it enables the vendor revenues to be at least two times higher.

Mind Genomics is based on sophisticated math, on the notion of value exchange on the top of fast Internet, and on DataMining. Mind Genomics is based on the following three production phases:

a. Micro Science (the phase based on sophisticated math).
b. Customer Typing (the phase based on value exchange and fast Internet).
c. Machine Learning (the phase based on DataMining).

In this context, sophisticated math implies linear regression and related techniques for improving the effects of linear regression. Value exchange implies that whenever a customer is asked to do something, he/she must receive an immediate award. DataMining spans all techniques that can help the system self-learn and adapt to the changing marketing requirements.

1.1 Phases of Mind Genomics

In essence, in the first phase, Mind Genomics determines, for a given product or service, in a given geographical region, what the mind types of potential customers (prospects) are; in other words, in the first phase, Mind Genomics determines what the major customer types are, i.e. the "baskets" at which incoming customers are to be placed upon arrival to the web portal of the vendor. As it will be seen later on, based on the empirically obtained experience, Mind Genomics finds that it is enough to create only three "baskets" i.e. it determines only three customer types, occasionally four.

In the second phase, Mind Genomics waits for a customer to arrive to the web site of the vendor, serves his/her needs, and then, after all customer needs are served properly, offers a value exchange. In this context, value exchange means that the customer is asked to answer three multiple-choice questions, each on a simple 3-point scale, and in return, the customer is offered a compensation for the unexpected effort (the three additional clicks represent an additional effort which is typically not done without appropriate compensation). The type of compensation used depends on the type of the vendor business. For example, in the case of restaurant business, a free appetizer could be offered (one of several to choose from); in the case of vet clinics, a utility for pets could be offered.

In the third phase, techniques from machine learning in general and DataMining in particular are used, in order to improve the operation and to make the system more effective.

Mind Genomics can also be treated as a method to implement horizontal segmentation and horizontal segmentation, to be treated later, can be treated as a method to improve market effectiveness. Mind Genomics can be used not only in retail marketing, but also in education, science, or politics, to promote specific new products, services, ideas, opportunities, etc.… Mind Genomics combines the general (in the first phase), the specific (in the second phase), and look-ahead (in the third phase).

Mind Genomics, in its first phase, brings the customers to the business (to the web portal of a business or to the physical door of a business); in its second phase, maximizes the revenues in a way that makes the customer happy, so he/she returns many more times. In its third phase, which is still in development, the system gets improved. Since the third phase is still in development, it will not be further elaborated here.

The first phase consists of four sub-phases (stages):

a. Building of the MicroScience
b. Polling of the market
c. Analysis of the poling results
d. Selection of optimal marketing slogans.

The second phase also consists of four sub-phases (stages):

a. Typing of the customer
b. Determination of the target action
c. Analysis of the profit
d. Building the experience base.

1.1.1 Phase #1: Building of the MicroScience (Stage #1)

In this stage of phase #1, one has to create silos (empirical studies indicate that six silos is optimal), slogans (empirical studies indicate that 36 slogans is optimal, six per silo), and demographic specifiers for pollees to be polled (empirical studies indicate that 300 pollees is optimal). Figure 1.1 gives an example of some silos and ideas (slogans) within the silos.

Previously, it was indicated that Mind Genomics does not use demographics. Actually, it uses demographics in the first phase, the MicroScience, when the goal is to determine numerically, as it will be seen later, of the initial 36 slogans, what three slogans are the best suited, for the given product or service, in a given geographical region.

A silo is a domain of interest for the marketing campaign. For example, in restaurant business, possible silos could be: (a) food, (b) wines, (c) music type, (d) ambient atmosphere (e) the waiters' behavior, and (d) parking space. Another set of silos may also be appropriate for a given marketing campaign (see Fig. 1.1).

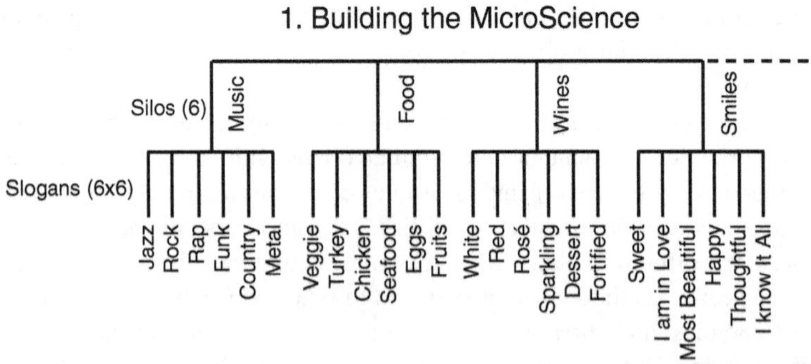

Fig. 1.1 Silos (6) and slogans (36)

Each slogan is supposed to consist of 3–4 elements: One element from the native silo and 2–3 more elements from another 2–3 silos. More specifically, if the native silo for a slogan is food, then, in the six slogans for which this silo is native, the first element should be about food, and the other two elements can be about wines, music, or anything else from the remaining five (5) silos.

An example of a 3-element slogan along these lines is: "Come to our restaurant to enjoy Mexican food, wines from Chile, and the tango music from Argentina."

The demographic specifier for polling should correspond to the characteristics of target customers. For example, when a restaurant is near a university, target customers could be students, staff, and professors. So, the demographic specifier may specify that 150 of the pollees should be male and 150 female; also, 200 young, 50 mid age, and 50 elderly; 200 low budget and 100 mid budget, etc.…

At the end of this stage, all the above data (6 silos, 36 slogans, and 300 specifiers) are forwarded to the Mind Genomics center for further processing. Note that silos, slogans, and specifiers are generated in one of many Mind Genomics office around the globe, while the Mind Genomics center at the Mind Genomics headquarters performs the more sophisticated stages of the entire process.

In principle, other organizational models are also possible, but this one is probably the most effective from the customer's point of view, and from the point of view of the protection of the intellectual rights of the technology creator and owner.

Once the data are received by the Mind Genomics center, a set of N vignettes is created. Typically, $N = 24–40$, but some studies reveal that in some cases 48 may be better suited. Anyway, from now on, it will be assumed in this text that $N = 40$.

A vignette is a screen with four of the 36 slogans. These slogans may not be semantically related to each other. A psychological theory says that, when a person sees a number of apparently unrelated sentences (a slogan is typically a sentence), based on the past life experiences, he/she fills in the missing facts, and creates a scenario with an opinion about the following issues:

(a) How much he/she likes the scenario?
(b) How much is he/she willing to pay for the realization of this scenario?

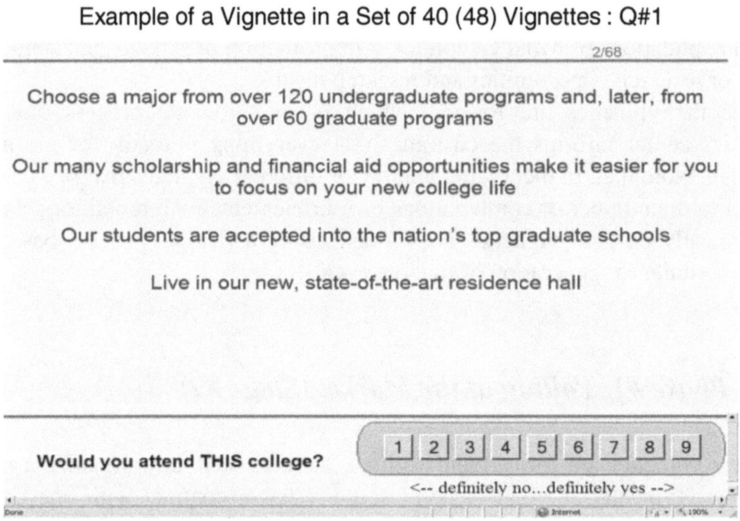

Fig. 1.2 An example of a vignette and a customer satisfaction related question

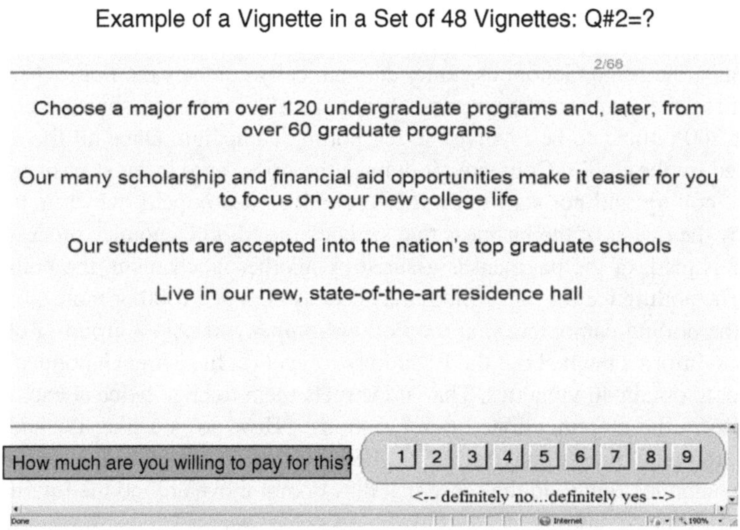

Fig. 1.3 An example of a vignette and a customer finances related question

Answers to the above questions are best given on a scale (from 0 to M), rather than in concrete terms (filling in descriptions in the case of the first question and dollar amounts in the case of the second question). One study (Moskowitz 2007) reveals that the optimal value for M is equal to 9, so the questions should be answered on the scale 1–9.

An example of a vignette is given in Fig. 1.2 (with the first question inserted) and in Fig. 1.3 (with the second question inserted). This example is related to promotion of a school program, on the college level. In general, one of the most fruitful applications of Mind Genomics is in promotion of schools and school programs, or research opportunities and research results.

Once the vignettes are formed, which is an automatic process, the Mind Genomics center informs the customer that everything is ready for the polling campaign. Note that, in theory, the number of different vignettes that one can form is equal to the number of combinations of four elements with repetitions. In practice, typically only 40 vignettes are formed, using a random process based on a precise formula for generation of random events.

1.1.2 Phase #1: Polling of the Market (Stage #2)

After the vignettes are formed and verified, the next step is to organize a polling campaign. Typically, a Mind Genomics center has a contract with one or more organizations that can do an effective polling campaign (PollingCompany). One of them is Amazon Turk. Among other things, it can organize a polling campaign. This company has millions of persons that signed up for participation in polling campaigns, in return for a benefit. This benefit can be in money, or in discounts on selected services of the Amazon system, or both combined.

Ideally, the Mind Genomics center automatically contacts the PollingCompany, announces the number of vignettes, their general theme, and shares the specifiers for 300 pollees to be involved in the polling campaign. Once all the above is obtained by the PollingCompany, it defines the price, and, in most business models, the polling will not start until after the polling fee is paid, which is typically done by the owner of the business that initiated the Mind Genomics process. Once the fee is paid, or the payment is assured by another mechanism, the polling can start. The polling fee for 300 pollees may cost as high as $1000 or more.

In the polling campaign, first the PollingCompany selects a group of 300 pollees, and informs them about the IP address to visit (at the Mind Genomics center) in order to obtain 40 vignettes, Than it instructs them to click twice at each one (in response to the aforementioned two questions: “How do you like it?” and “How much are you willing to pay for it?”). The vignettes do not leave the server of the Mind Genomics center, in the sense that they do not move around the Internet.

Next, at their convenience, each pollee (*a polled person*) visits the requested IP address, clicks 80 times to the 80 questions of the 40 vignettes, two rating questions per vignette. The responses of pollees have to be checked for consistency. One possible scenario is that a pollee automatically clicks 96 times without paying attention to the contents of questions, and then runs away to the next paid polling task. This scenario can be caught if control slogans are inserted into some of the vignettes. For example, some of the slogans could be inverted (instead of “I like white color" the system could insert "I do NOT like white color"). In this case, if the pollee was clicking without paying attention, his/her answers would include a contradiction, which is easy to catch. Of course, if a contradiction is recognized, the Mind Genomics will request a substitute polee. The PollingCompany will provide substitutes, as soon as it verifies the contradictions. The entire process can be fully automatized, and it actually is, in the case of Mind Genomics Advisors, LLC.

Finally, a three-dimensional matrix of polling results is ready and its analysis can start. Dimensions of the matrix of polling results are:

(a) The demographic dimension: 300 (pollees)
(b) The vignettes dimension: 40 (vignettes)
(c) The preference dimension: 2 (clicks on two questions per vignette).

All together, the matrix of polling results contains 24000 data items. In the next stage of the process, these data items are analyzed, in order to determine what are the major three mind types on the market (of a given product or service, in a given geographic or demographical region), and at the same time, what is the most effective marketing slogan (out of the initial 36) for each one of the major three mind types. Note that the goal is not to find the most effective three marketing slogans in general, but the three marketing slogans that are the most effective for the three determined mind types, one per type. These three specific slogans are rarely equal to the three generally most popular slogans.

1.1.3 Phase #1: Analysis of the Polling Results (Stage #3)

The first step of the analysis is to map the matrix of polling results into a two-dimensional space (X, Y) in which the X axis represents the scores (0–9) related to one of the two questions related to each vignette (e.g., "How you like it?"), and the Y axes represents the scores (again, 0–9) related to the other question (e.g., "How much are you willing to pay for it?"). One possible mapping scenario is shown in Fig. 1.4.

Each ball in Fig. 1.4 corresponds to one click in the polling campaign, and includes four slogans. Each slogan can be given a weight corresponding to the numerical values of the corresponding clicks. Such a setting opens doors for the application of a Bayes analysis. However, before that, the clusters of customers with similar mind types have to be determined. That is done using a form of linear regression, as indicated in Fig. 1.5.

Of the four clusters, one is considered less important and therefore is neglected. The other three are declared to correspond to the major three mind types on the market, and are marked as indicated in Fig. 1.6.

The next step is to determine what slogan is most effective for each mind type. One needs only to select the elements with the highest numerical values. If there is a compelling reason, one can choose other elements with high scores, and not

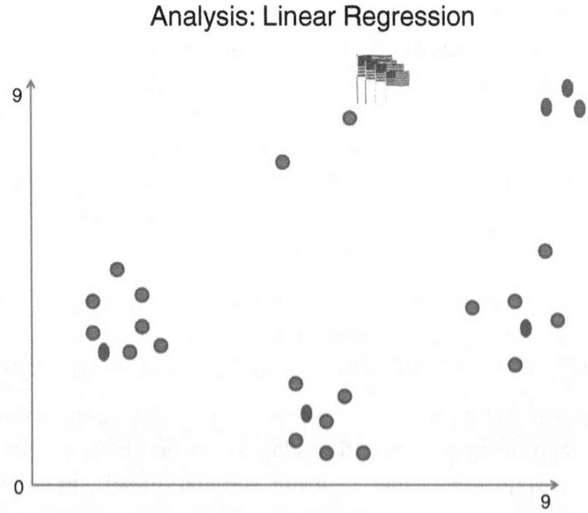

Fig. 1.4 One possible scenario for mapping the matrix of polling results into a two-dimensional space

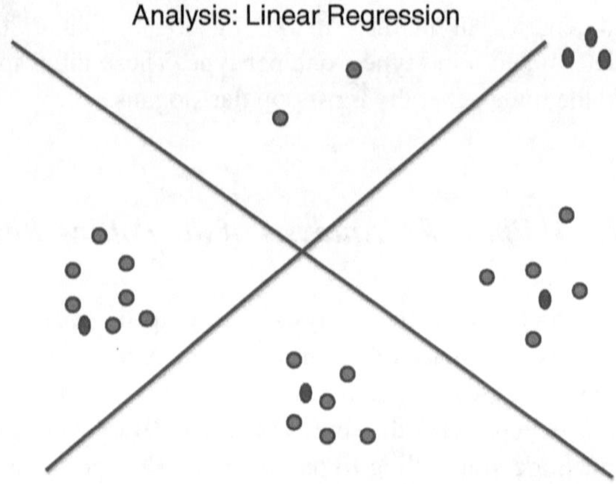

Fig. 1.5 The polling data divided onto clusters, with two lines, using a form of linear regression

the highest ones. One then creates a slogan with numerical value. In other words, a weighted sum is created, and the slogan with the highest weighted sum is declared to be the most effective slogan for the corresponding mind type. For example, Fig. 1.7 shows the major three slogans for the case of a product. In other words, *of the initial 36 slogans, the major three are not selected arbitrarily, but by a rigorous mathematical procedure!*

The story behind the three slogans in Fig. 1.7 could be attributed to sales of shrimps after the oil spill in the Mexican gulf, and could be interpreted as follows:

(a) The slogan #1 is oriented to patriotic mind types and could read as follows: "Better buy shrimp from a USA company. You know how it was contaminated, you also know that it was decontaminated using a superior technology

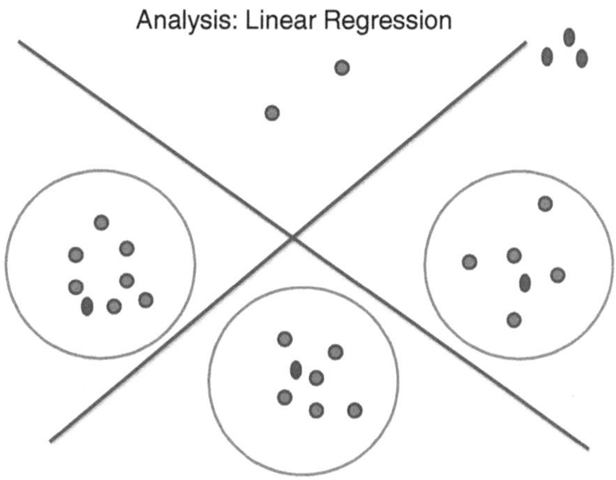

Fig. 1.6 The regression analysis determines the three clusters corresponding to the major three mind types on the market

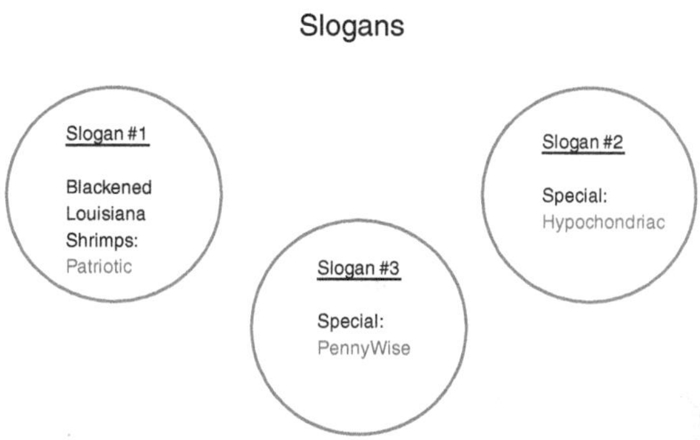

Fig. 1.7 Three example slogans, selected out of the initial 36

of the USA. Buy this American shrimp rather than buying shrimp from another country, for which you think that it was not contaminated, but maybe it was contaminated with bacteria and parasites not yet known about—and if decontaminated at all, it was with a technology which is inferior in comparison with the technology of the USA."

(b) The slogan #2 is oriented to stingy mind types and could read as follows: "Our shrimp is the cheapest in the nation. Period!"

(c) The slogan #3 is oriented to hypochondriacs and could read as follows: "During the decontamination process, our shrimp was enriched with the strongest anti-cancer materials."

In that same way one can develop a number of different scenarios. For example, if the system is set to promote collateral sales around a small veterinary clinic for home pets, then the selected set of the most effective slogans may read as follows:

(a) Slogan #1, for lonely people:
"You are welcome to subscribe to our "Lonely Hearts Club Band".
(b) Slogan #2, for dandy people, who like to show off:
"We have Armani and Versace coats for pets, which you can select when you bring your pet"
(c) Slogan #3, for animal lovers:
"We have other animals that you can purchase, when you bring your dog for examination."

Many other examples could be designed, which can be an exercise for students and other interested readers. Actually, the entire process explained till this point is treated as a homework assignment or an issue for scientific discussions.

1.1.4 Phase #1: The Broadband Marketing Campaign with the Best Three Slogans (Stage #4)

The essential point of the technology presented so far is to do the selection of the best three slogans using a scientific method, rather than doing it arbitrarily. An analysis has shown (Moskowitz 2007) that the scientific approach generates about 30 % more interest among prospects (potential buyers), compared to the arbitrary approach.

Once the three slogans are determined, the next step is to launch them into the market, via media like: Radio or TV, Internet or intranet, billboards or taxis. No matter what media are used, the messages will reach the prospects, and some of them will visit the web site of the party (company or clinic or school) that initiated the Mind Genomics campaign. Once they reach the web site of the selling party, the Micro Science phase ends (Phase #1) and the Customer Typing phase can start (Phase #2).

1.1.5 Phase #2: Typing of the Customer (Stage #1)

So far it was assumed that the interested prospect will visit the web site of the selling party, and that, after he/she finishes with the planned task (reserving a seat in a restaurant or reserving an examination slot in a pet clinic), he/she will be offered a value exchange (to click three times on three triple-choice questions and to get a free gift in return). However, in general, a customer can just bump into a restaurant or walk into a clinic, without a previous slot reservation activity.

In the Internet scenario, the customer typing is done via the Internet, after the customer finished the initially planned activity. In the walk in scenario, the customer satisfaction could be done after walk in, by offering the just-walked-in customer to take a tablet PC, and to click three times to three triple-choice questions

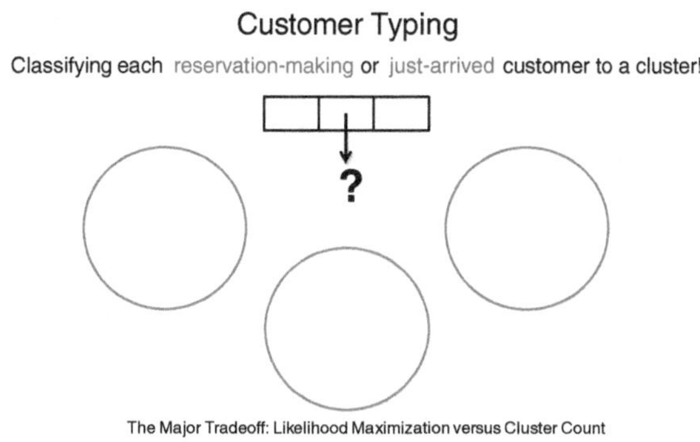

Fig. 1.8 Three triple-choice questions are being asked. In general, the customer type can be determined after only one triple-choice question is asked. However, the system contains "noise" and after the first question is answered, the system assigns a guess and assigns a probability that the guess is correct, as indicated in Fig. 1.9.

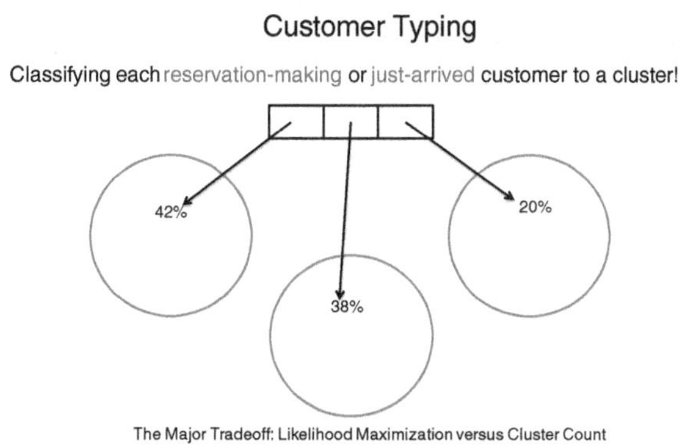

Fig. 1.9 After the first triple-choice question is answered, the system creates a guess

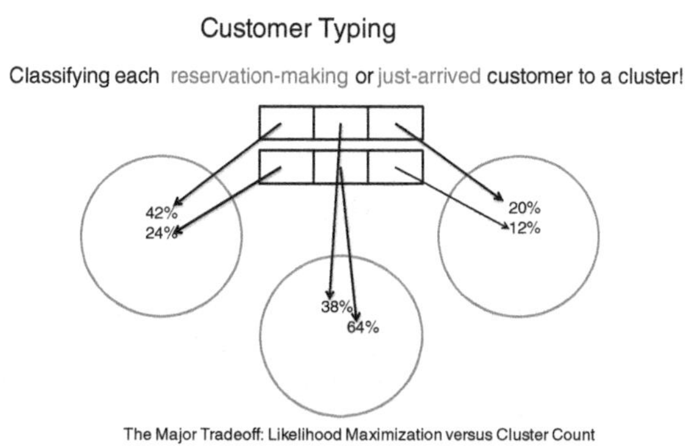

Fig. 1.10 After the second question is answered, the system likelihood to guess correctly has increased

which will allow the owner of the business to serve the customer better. In any of the above two cases, the presented question may look as in Fig. 1.8.

The guess may be altered after the second question is answered, as indicated in Fig. 1.10, while the system is 88 % sure that the guess is correct, after the third question is answered, as indicated in Fig. 1.11. The technicalities related to the

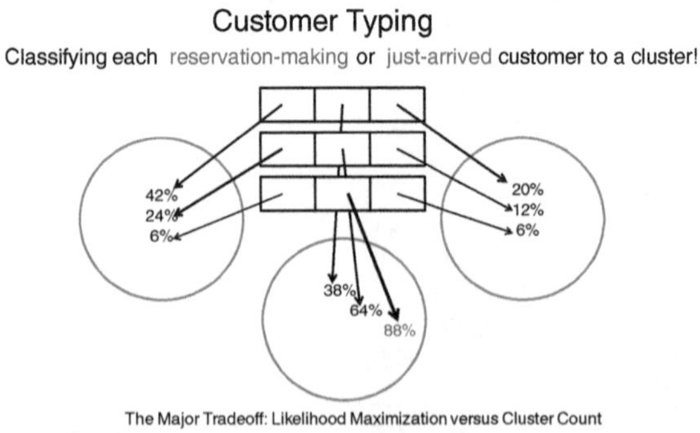

Fig. 1.11 After the third question is answered, the system guess of the customer mind type is correct with relatively high probability

generation of questions and the forming of the guess probability have been developed over several years, and reside both in statistical know-how, and in patented algorithms.

1.1.6 Phase #2: Determination of the Target Action (Stage #2)

Once the customer type is determined, the "magic" sentence is passed to the customer. In this context, "magic sentence" means a text which is likely to motivate the customer to the purchase, the one that he/she initially was considering, or a follow up purchase. A study reveals (Moskowitz 2007) that Mind Genomics enables that sales increase as much as 160 %. The typical increases are 15–50 %. Of course, the assumption is that increased purchases increase the customer satisfaction, so the customer is happy and comes back again.

1.1.7 Phase #2: Analysis of the Profit (Stage #3)

The best way to analyze the profit generated by a novel marketing method is to own two or more sales units, and to apply the new marketing method only to one sales point or a subset there-off. In that case, the differences in effectiveness are easily seen. This is called an A/B test. Of course, they are not necessarily seen immediately, but perhaps after a time period of 1 year. This lag occurs because sometimes the sales increase also due to other factors, not related to messages, such as the change of season (the fall sales may be better than the summer sales).

1.1.8 Phase #2: Building the Experience Base (Stage #4)

Once the system is exploited for some time, the user acquires some experiences and gets ideas on how to improve it. This part leads to new research efforts. In the context of this book, students and interested readers are asked to provide ideas on how to improve the technology of Mind Genomics, in order to improve its effectiveness.

1.1.9 MicroScience Revisited (Phase #1 Revisited)

In the previous text, it was clearly explained what the steps to be done in the Micro Science phase are, but it may not be clear enough, who does these steps. The "who" issue should be fully understood from Figs. 1.12, 1.13, 1.14, 1.15, 1.16, 1.17, 1.18, 1.19, 1.20, 1.21, 1.22, 1.23 and 1.24, in the case of Micro Science. The figures were set in a way that specifies the chronology of the process. In the figures to follow, not only 1.12 to 1.24, but also later, in the Figs. 1.25 to 1.37, the abbreviations have the following meanings:

(a) POLLEES: Individuals involved in the polling campaign
(b) WEST [STANFORD]: The PollingCompany [SU = Stanford University]
    (In a student case study it can be assumed that it is located in the West of USA, a spun off SU)
(c) EAST [HARVARD]: The Global Mind Genomics Center
    (In a student case study it can be underlined that the ideas spawned off Harvard University)
(d) MEDIA[30] = The media used in advertising
    (Students can be asked to develop Mind Genomics campaign for 30 different media)
(e) CUSTOMER = a prospect
(f) REST = RESTAURANT
    (In a student case study a restaurant example could be used for easy understanding)
(g) MGS/S = A local Mind Genomics Center in a given state
    (This is where the parties interested to sell using Mind Genomics report to)
(h) PSYHO/AN = A psycho-analyst employed
    with a local Mind Genomics center (That person is responsible for creation of slogans).

The process which flows from Figs. 1.12, 1.13, 1.14, 1.15, 1.16, 1.17, 1.18, 1.19, 1.20, 1.21, 1.22, 1.23 and 1.24 involves activities (built into arcs of the graph) which are meant to be self-explanatory. Students and interested readers could be asked to explain precisely the meaning of all abbreviations in the arcs, as an exercise. In general, the process can be treated as machine learning.

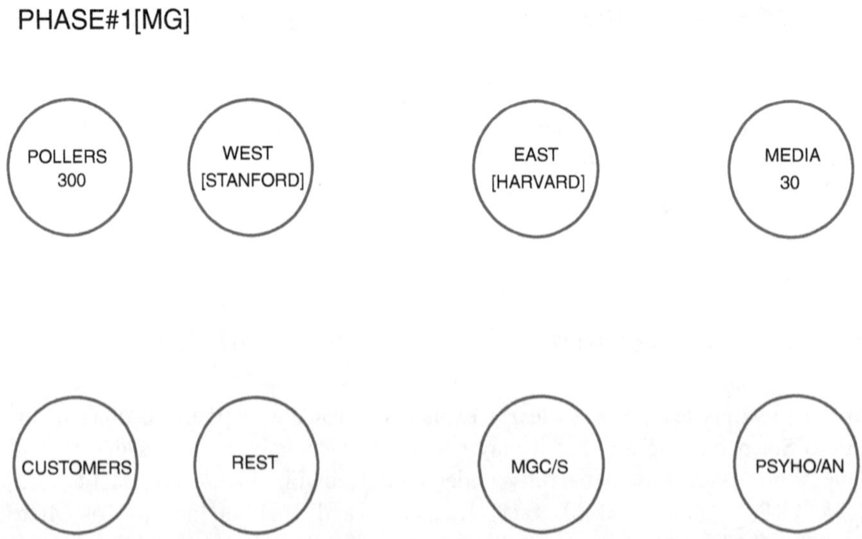

Fig. 1.12 *Phase #0* This figure indicates who the major participants in the process are Explanations are given in the section on MicroScience

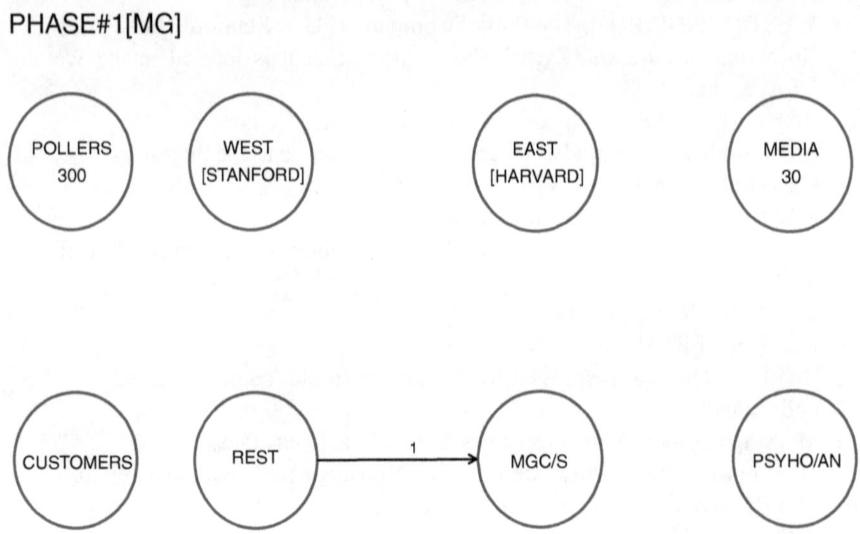

Fig. 1.13 *Phase #1/Action#1* The restaurant owner visits the Mind Genomics center and places the request for a Mind Genomics campaign

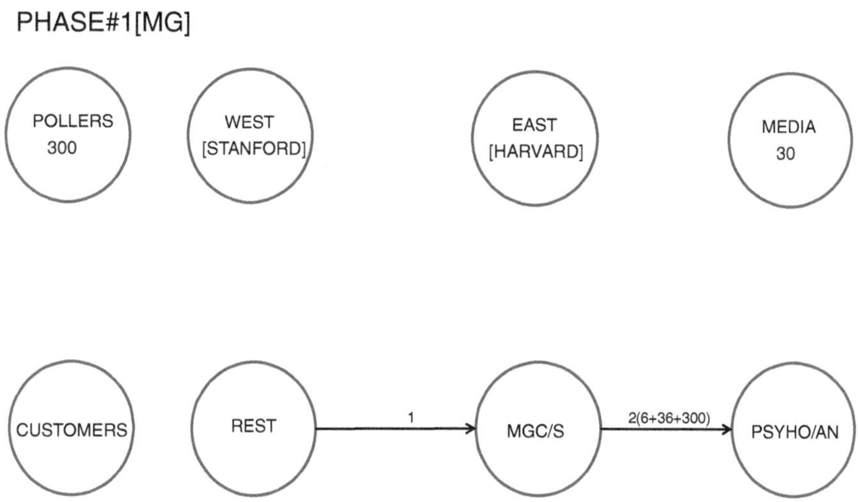

Fig. 1.14 *Phase #1/Action #2* The Mind Genomics center, through the help of its psychoanalyst, creates 6 Silos, 36 Slogans, plus demographic specifiers for 300 pollees

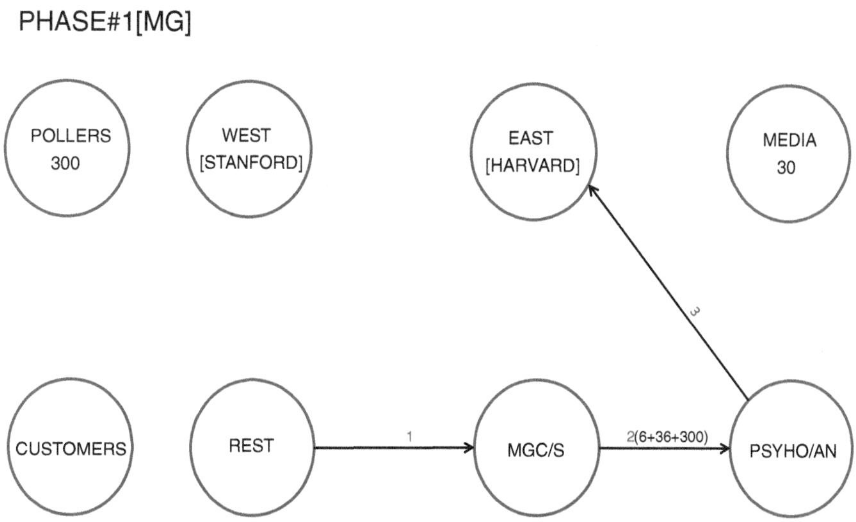

Fig. 1.15 *Phase #1/Action #3* The generated materials (Silos, Slogans, and demographic specifiers) are submitted to the Mind Genomics headquarters

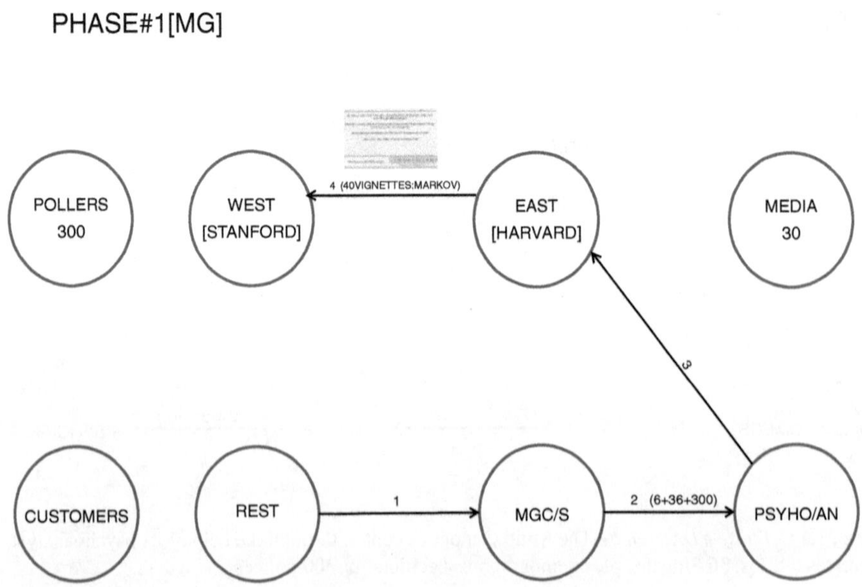

Fig. 1.16 *Phase #1/Action #4* The vignettes (typically 40 or 48) are generated at the Mind Genomics headquarters; then, the poller is informed that the vignettes are ready and that the pollees are welcome to access the vignettes at the server in the Mind Genomics headquarter

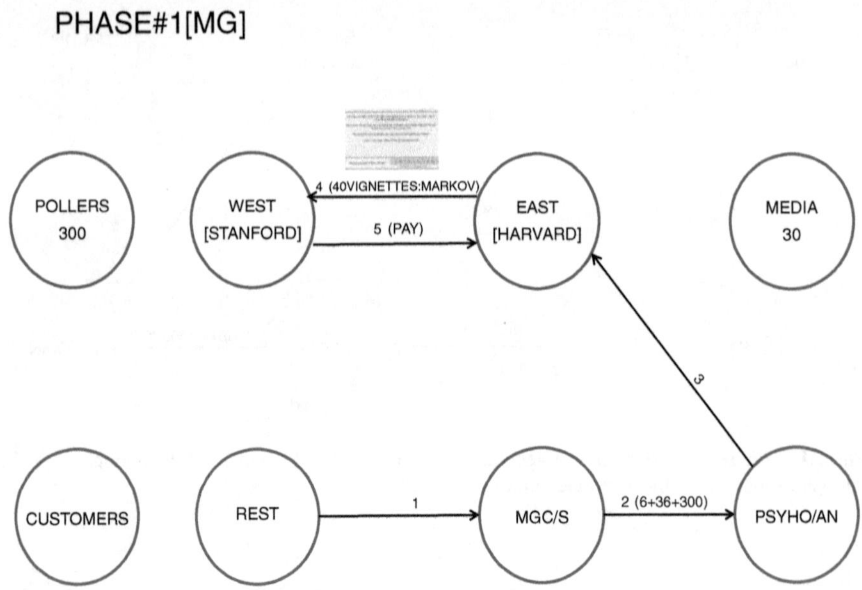

Fig. 1.17 *Phase #1/Action #5* The poller agency informs the Mind Genomics headquarter that the poll will start as soon as the polling services are paid in advance

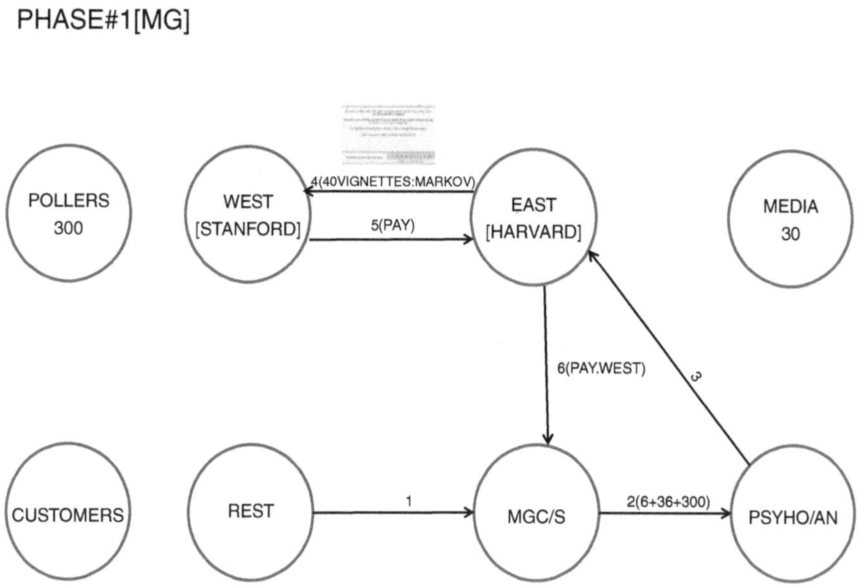

Fig. 1.18 *Phase #1/Action #6* The request for the payment is forwarded to the local Mind Genomics center

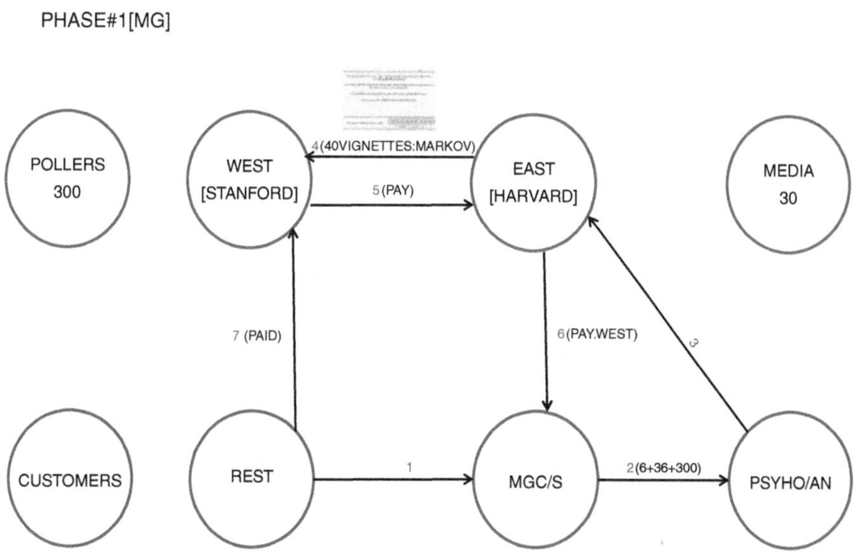

Fig. 1.19 *Phase #1/Action #7* The restaurant owner pays the poller

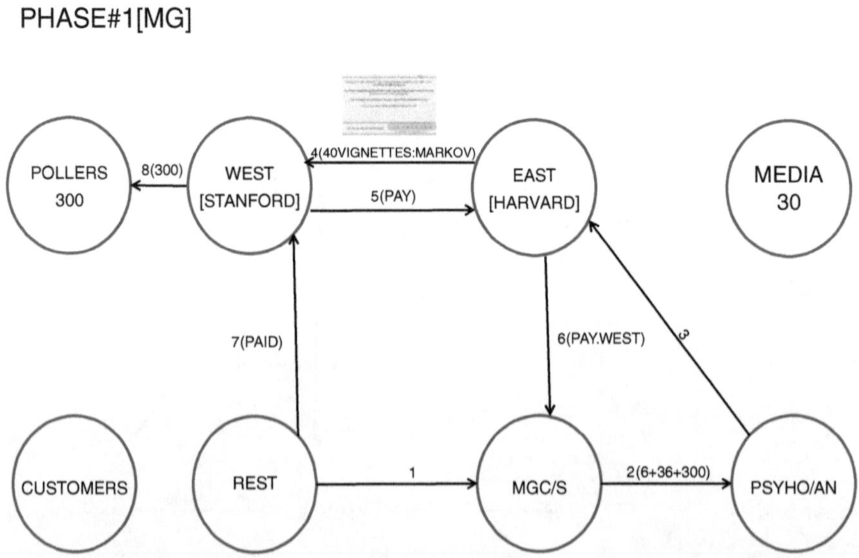

Fig. 1.20 *Phase #1/Action #8* The poller triggers the pollees to start the polling process

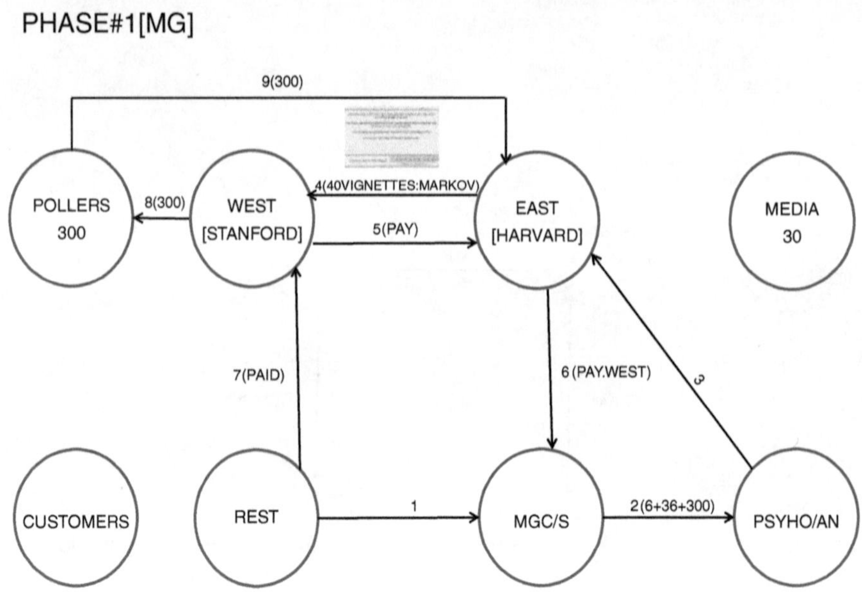

Fig. 1.21 *Phase #1/Action #9* The pollees access the vignettes (each pollee clicks 80 or 96 times, twice per vignette)

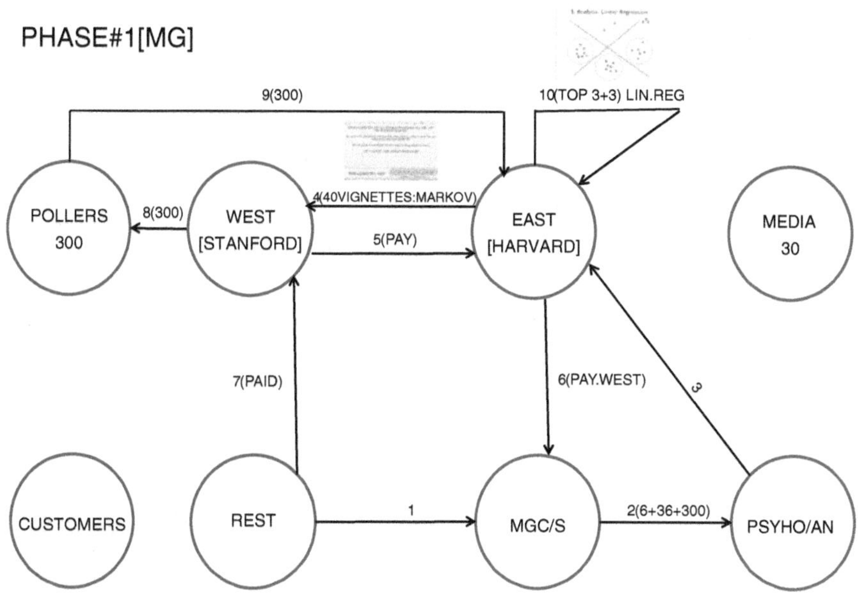

Fig. 1.22 *Phase #1/Action #10* Using linear regression, the mayor three mind types are determined, as well as the most effective slogans for each one of these three mind types; all this is done at the Mind Genomics headquarters

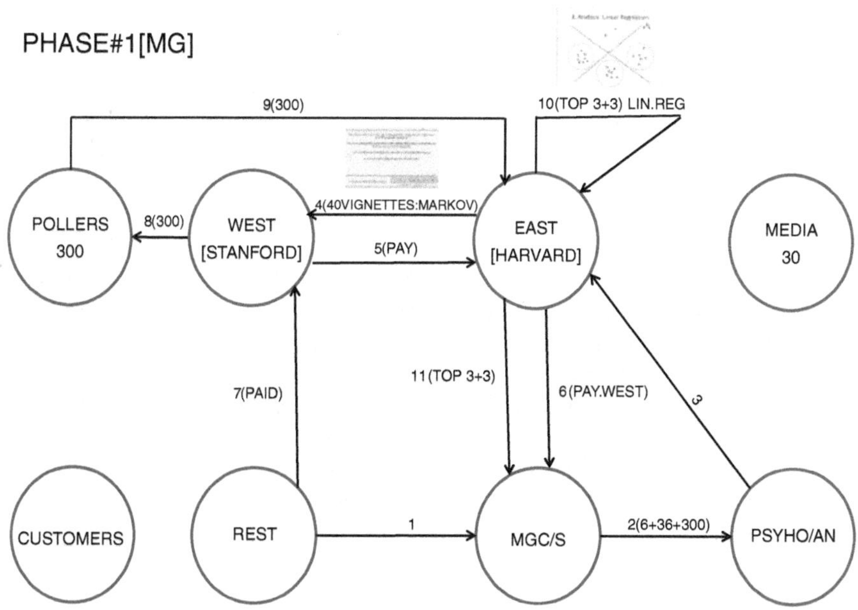

Fig. 1.23 *Phase #1/Action #11* The information about the mind types and the slogans is passed to the local Mind Genomics center

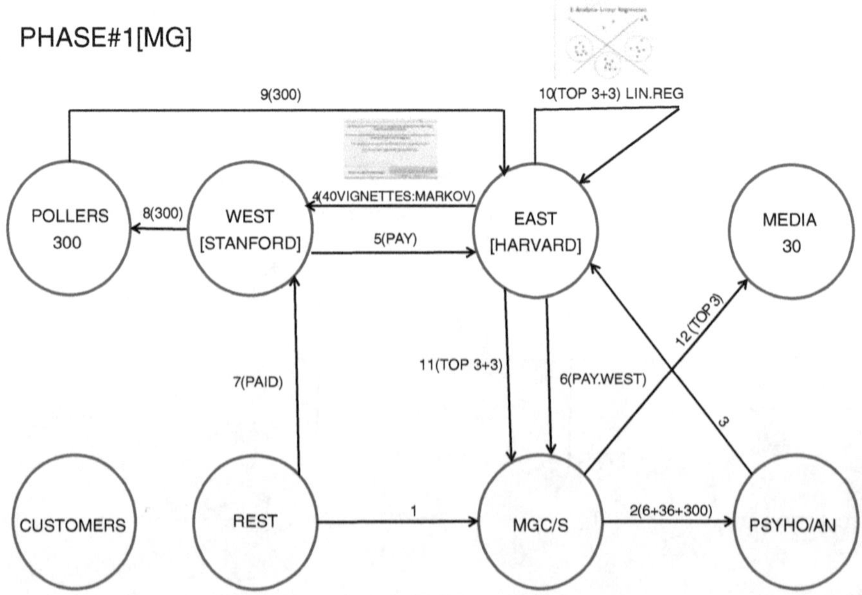

Fig. 1.24 *Phase #1/Action #12* The top three slogans are forwarded to media

1.1.10 Customer Typing Revisited (Phase #2 Revisited)

Again, in the previous text, it was clearly explained what are the steps to be done, but it may not be clear enough, who does these steps. The "who" issue should be fully understood from Figs. 1.25, 1.26, 1.27, 1.28, 1.29, 1.30, 1.31, 1.32, 1.33, 1.34, 1.35, 1.36 and 1.37, for the case of Customer Typing. The figures were set to reveal the chronology of the process. As before, the process which flows from Figs. 1.25, 1.26, 1.27, 1.28, 1.29, 1.30, 1.31, 1.32, 1.33, 1.34, 1.35, 1.36 and 1.37 involves activities (built into the arcs of the graph) which are meant to be self-explanatory. Students and interested readers could be asked to explain precisely the meaning of all abbreviations in the arcs, as an exercise.

The bottom line of Phase #1 is that mathematics and statistics are used rather than intuition and feeling, while in Phase #2 value exchange free selection are used rather than filling the forms and tracking the behavior. Unlike in modern targeted marketing, where demographic data are used both in customer profiling and in customer typing, Mind Genomics uses demographic data only in Micro Science, not in Customer Typing.

Fig. 1.25 *Phase #2/Action #0* The same set of players remains in the second phase

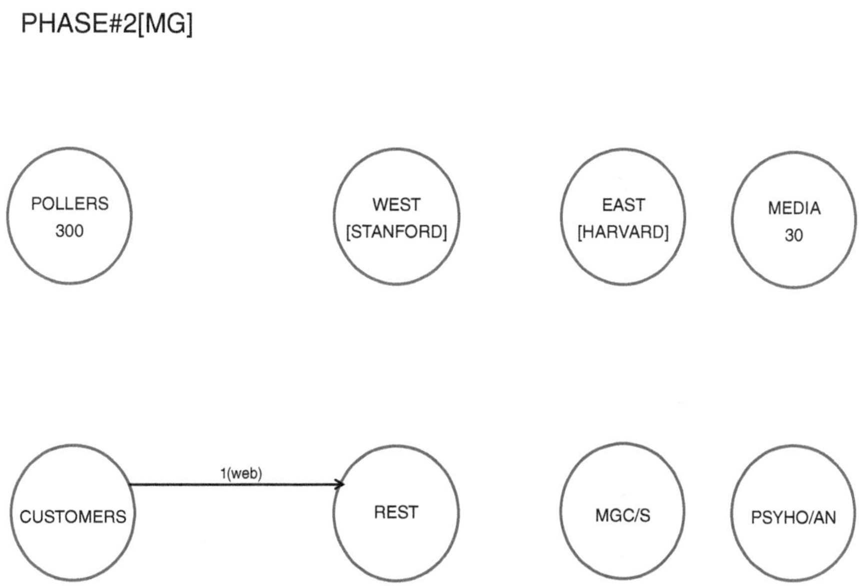

Fig. 1.26 *Phase #2/Action #1* A customer logs onto the restaurant's web site to book the table

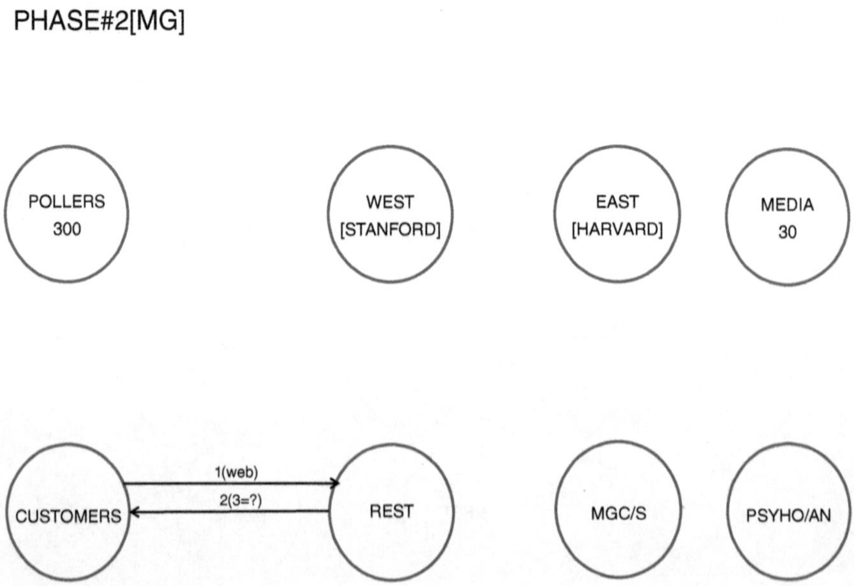

Fig. 1.27 *Phase #2/Action #2* The restaurant fulfills the customer's request and offers him/her a value exchange—to answer three more questions in exchange for a value (e.g., a free appetizer upon arrival at the restaurant)

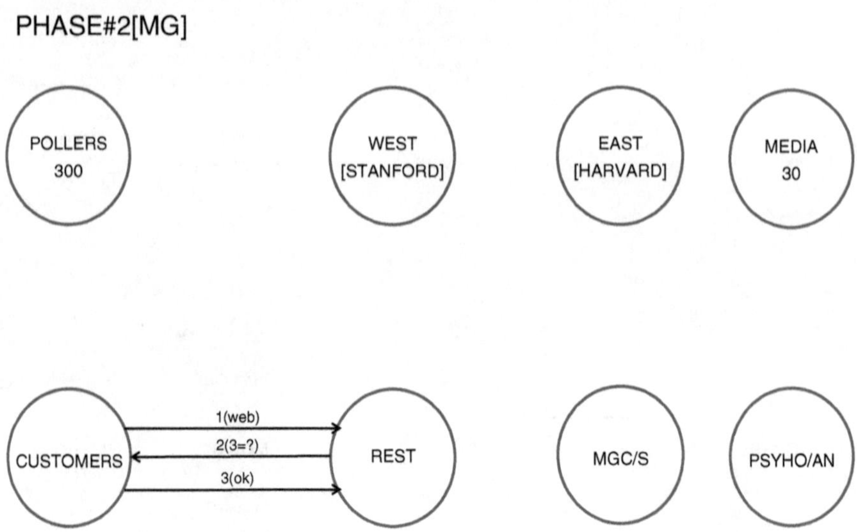

Fig. 1.28 *Phase #2/Action #3* Customer accepts the offered value exchange

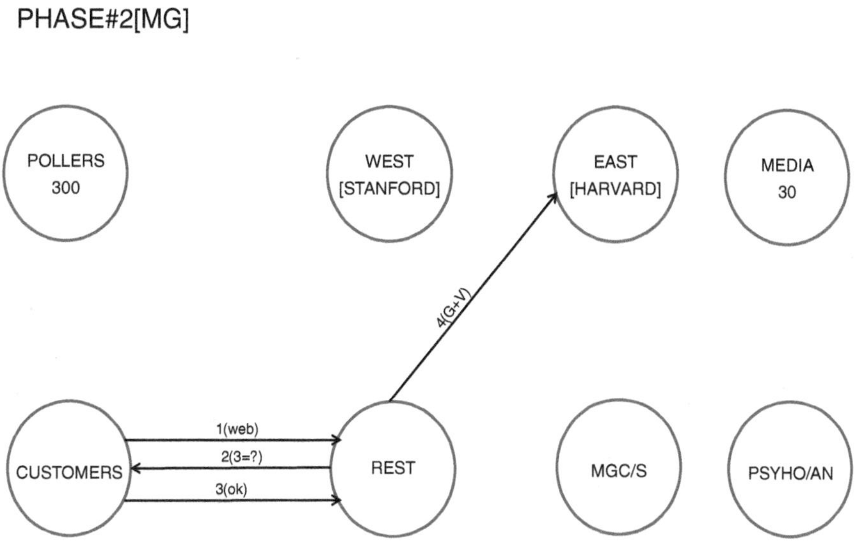

Fig. 1.29 *Phase #2/Action #4* The restaurant's server contacts the global Mind Genomics center, and asks for three questions to be passed to the customer on hold (obviously, Mind Genomics relies on fast Internet, because customers can typically wait only around 200 ms before they run away)

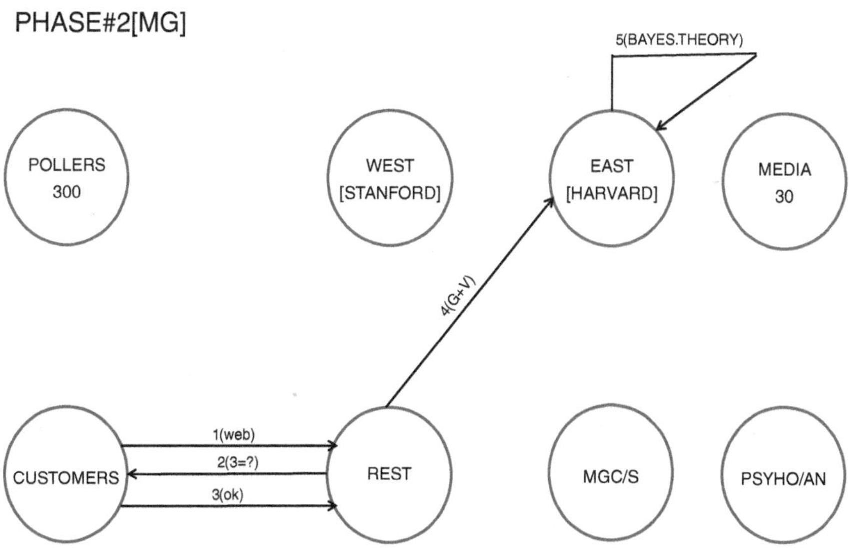

Fig. 1.30 *Phase #2/Action #5* Using the Bayes theory, the global Mind Genomics center generates the questions

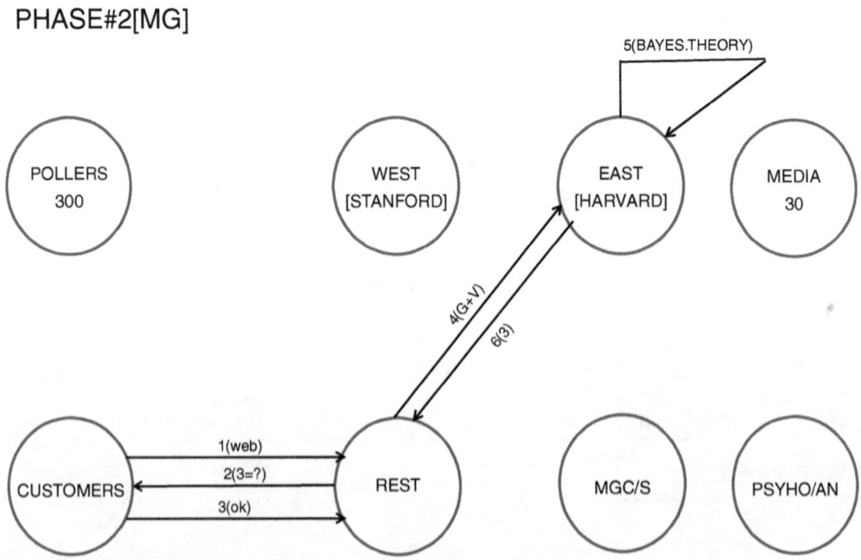

Fig. 1.31 *Phase #2/Action #6* The questions are passed to the restaurant's server

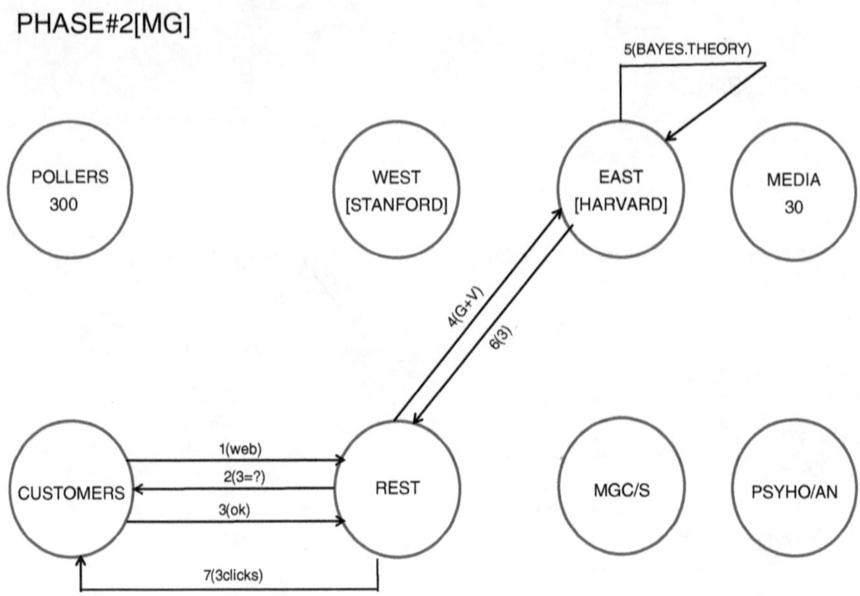

Fig. 1.32 *Phase #2/Action #7* The questions are forwarded to the customer on hold (hopefully in less than 200 ms)

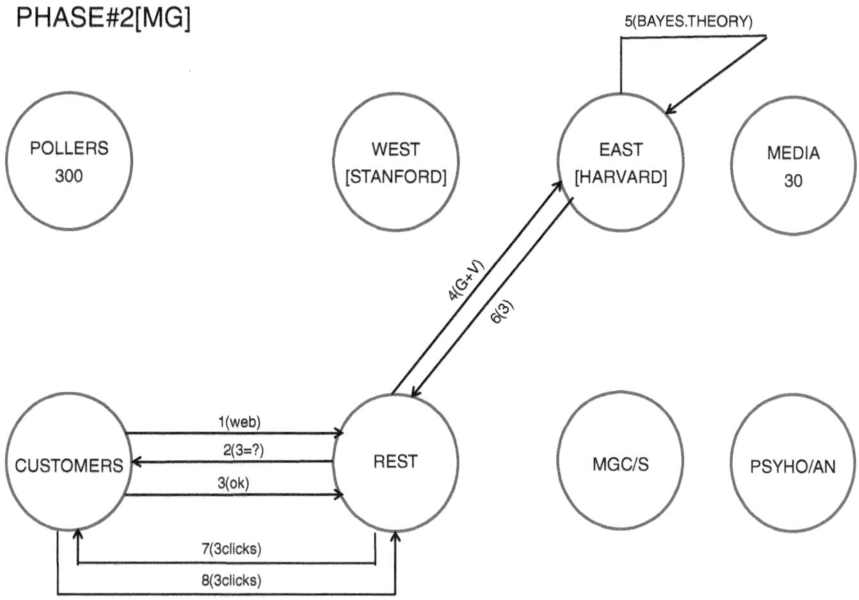

Fig. 1.33 *Phase #2/Action #8* The customer clicks three times

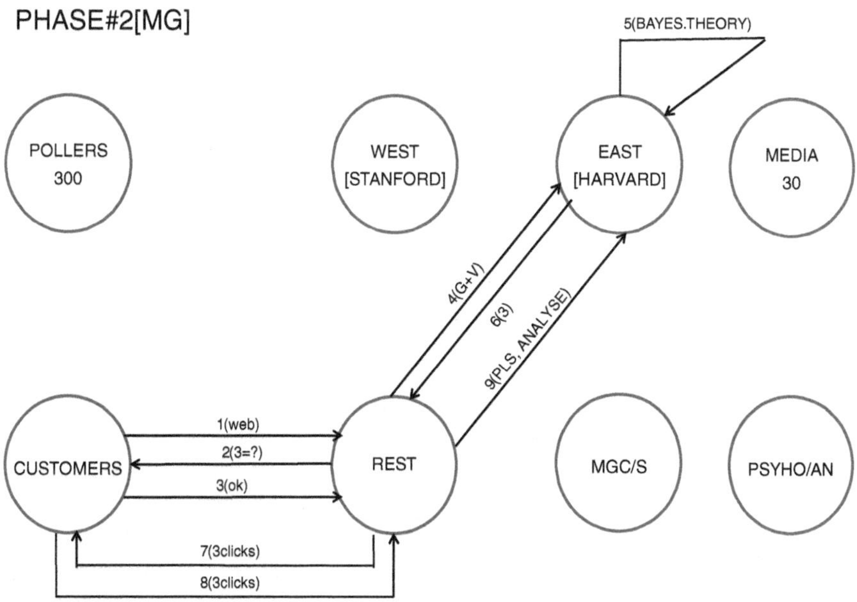

Fig. 1.34 *Phase #2/Action #9* The customer's answers are forwarded to Mind Genomics headquarters

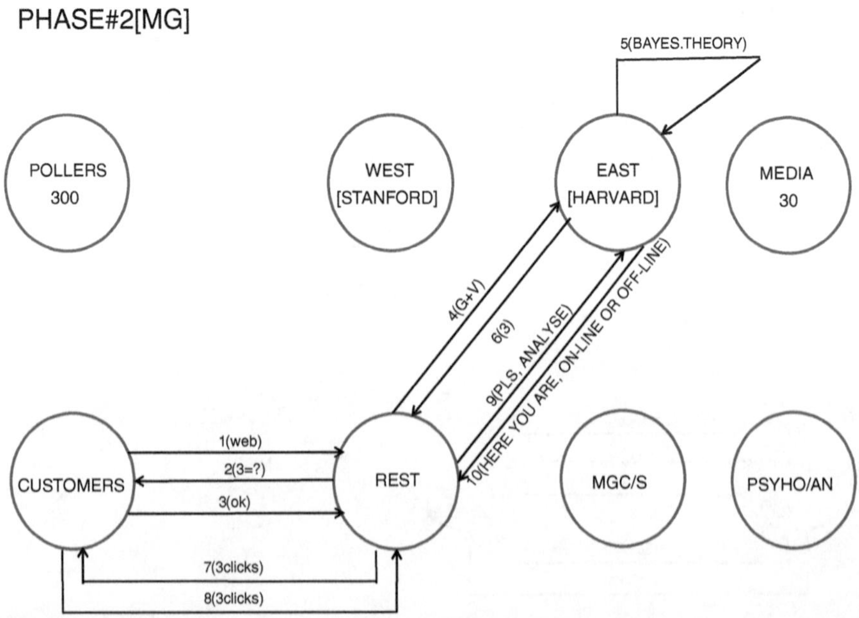

Fig. 1.35 *Phase #2/Action #10* The Mind Genomics headquarters determines the mind type of this particular customer and generates the most effective statements to be passed to the customer in order to maximize his/her spending in a way that makes him/her happy, so that he/she comes back again with a high probability

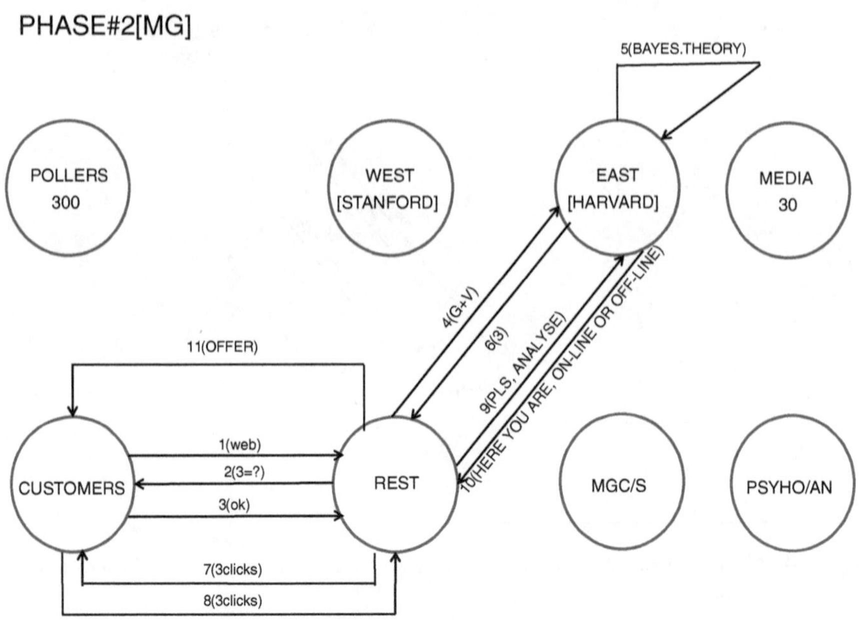

Fig. 1.36 *Phase #2/Action #11* The "magic statement" is passed to the customer

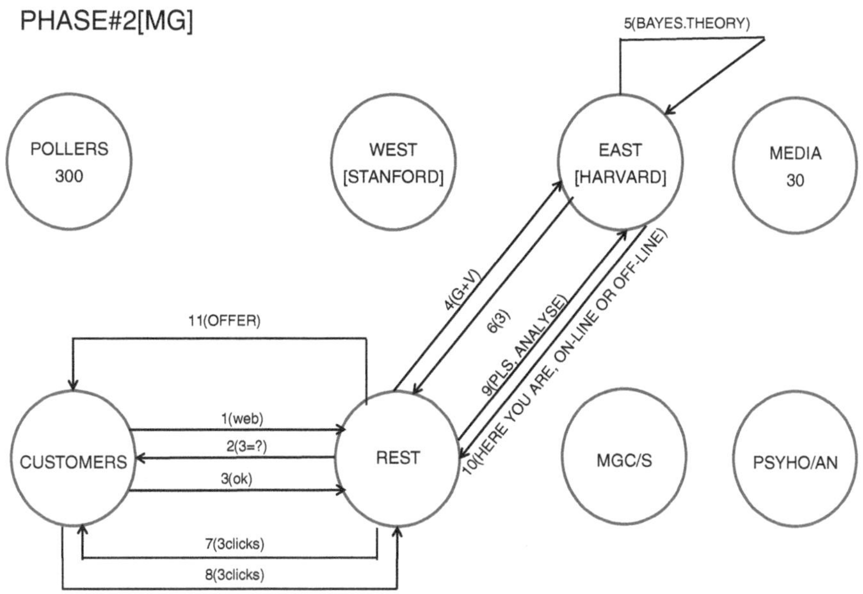

Fig. 1.37 *Phase #2/Action #12* The customer is happy

1.2 Horizontal Segmentation

All the discussions so far were presented for the case when the level of horizontal segmentation is equal to one. Since Mind Genomics can be used to support the practice of horizontal segmentation larger than one, the basic definition of horizontal segmentation will be discussed next, using Fig. 1.38.

The essence of horizontal segmentation is as follows: Assume that before horizontal segmentation is done, only 1000 pieces of a product (e.g., yogurt) could be sold per day, and for the maximal price of $1. After horizontal segmentation is done, maybe 4000 pieces of the same product could be sold, for a unit price of $2. Sounds like miracle, but it is not. Actually, the product is enhanced with a vector of added values. For example, some people like yogurt with fruit, and, to satisfy their need for fruit, they are willing to pay $2. Other people like yogurt with seeds, to help them in gym exercising between work and home. Still others may like yogurt with some anticancer materials to fight against AFLA toxin in yogurt. The reader can think of other good examples.

It is obvious that Mind Genomics can be used to determine what an added value that people like to have in yogurt is (Phase #1 of Mind Genomics). Also, once a customer logs onto the web site of a supermarket, Mind Genomics can be used to type the customer and to offer him/her the type of yogurt that he/she is not likely to refuse to buy (Phase #2 of Mind Genomics). If a person appears at the physical supermarket, someone from the welcome desk can type that person, by

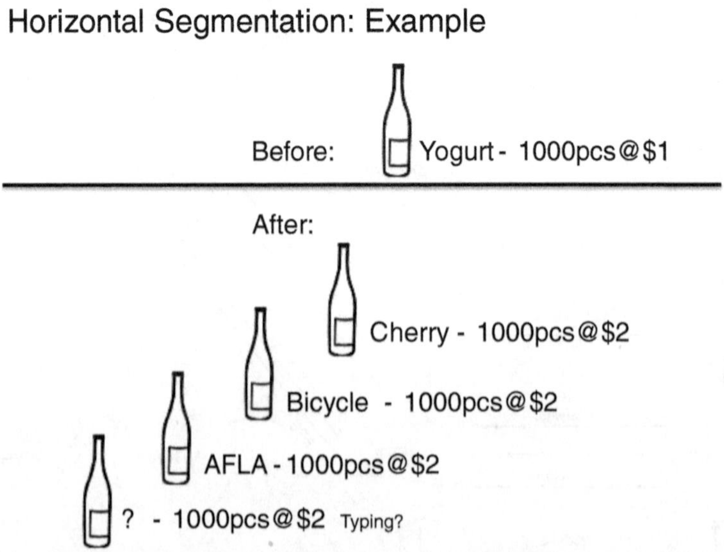

Fig. 1.38 Horizontal segmentation—a definition based on an example

asking him/her to click three times onto a tablet PC, in exchange for some value (as typical for Phase #2 of Mind Genomics). Of course, if nobody types the customer, he/she will get self-typed when he/she drops a bottle of yogurt into a shopping cart.

1.3 Definition of Some Terms and a Conclusion

Idea Map is a program that implements the processes that make Mind Genomics. Inputs for this program are: (a) Silos, (b) Slogans, and (c) Specifiers.

Questions in the Mind Genomics process are never demographic. In general, they belong to one of the following three categories:

(a) PSYCHOGRAPHIC ("What do you like in this photo?"),
(b) BEHAVIORISTIC ("Do you prefer formal or casual?"), or
(c) MERCANTILISTIC ("How much are you willing to pay?").

Addressable minds are the people (potential customers) who belong to one of the categories discovered. Only such people can be typed.

Some of the papers that impacted the creation of this book are given in Fig. 1.39. They are mostly related to the technology underneath the concepts of importance for B2C (Business to Customer).

Relevant Papers

Milutinovic, V., A Comparison of Suboptimal Detection Algorithms Applied to the Additive Mix of Orthogonal Sinusoidal Signals, IEEE Transactions on Communicatiions, Vol. COM-36, No. 5, May 1988, pp. 538-543.

Milutinovic, V., Knezevic, P., Radunovic, B., Casselman, S., Schewel, J., Obelix Searches Internet Using Customer Data, IEEE COMPUTER, July 2000 (impact factor 2.205/2010).

Milutinovic, V., Cvetkovic, D., Mirkovic, J., Genetic Search Based on Multiple Mutation Approaches, IEEE COMPUTER, November 2000 (impact factor 2.205/2010).

Milutinovic, V., Ngom, A., Stojmenovic, I., STRIP --- A Strip Based Neural Network Growth Algorithm for Learning Multiple-Valued Functions, IEEE TRANSACTIONS ON NEURAL NETWORKS, March 2001, Vol.12, No.2, pp. 212-227 (impact factor 2.889/2010).

Fig. 1.39 Some papers of relevance for implementation of the B2C infrastructure

In conclusion, the era of Mind Genomics is yet to come, since the enabler technology (fast Internet) is getting stronger and stronger, while the number of its users keeps growing more and more.

Reference

Moskowitz H (2007) Mind Genomics. In: Proceedings of the IPSI-2007. Harvard Club, New York, pp M.1–10

Chapter 2
Research Challenges

Abstract This chapter sheds light on various research areas of interest for further development of Mind Genomics. Research areas discussed refer to those of interest for the phase ONE of the process (avoiding the relatively high expenses related to market polling) and for the phase TWO of the process (avoiding relatively high expenses related to the estimation of the customer satisfaction level). Research of interest for the first phase is in media mining, while the research of interest for the second phase is in image understanding.

Keyword Mind genomics · Market polling · Estimation of customer satisfaction · Media mining · Image understanding for marketing purposes

From the financial point of view, the major research challenges for researchers in text and image understanding (TIU) and computer science and engineering (CSE) are related to the following two problems:

(a) How to avoid the polling expenses, in the micro science phase?
(b) How to avoid the value exchange, in the customer typing phase?

The first expense could be avoided when customer types are formed under the impact of media, like newspapers, magazines, radio, musical media, TV, or the Internet. The tools and methods from NLP (natural language processing), speech understanding, or image understanding, can help determine what the basic customer types are, and may avoid the payment to polling companies.

The second expense can be avoided when the customer typing is performed using a sentiment detection procedure based on text (via social networks), on voice (via telecoms), or on face understanding (via cameras), respectively.

In the next two subsections of this book, we briefly review the ongoing research in the above mentioned two subjects, and we direct the interested readers to the open literature that we reference. More on the subject will be included into a follow-up book.

V. Milutinovic and J. Salom, *Mind Genomics*, SpringerBriefs in Business,
DOI 10.1007/978-3-319-39733-7_2

2.1 Media Understanding Oriented Research

This section gives an overview of the ongoing research in media understanding, which spans from text understanding and still image understanding, all the way to understanding of moving image, and combinations of all the above. Each research effort is covered with one paragraph and one reference.

2.1.1 Understanding Text

In the article Fan et al. (2006), the authors give an excellent overview of the technology foundations of text mining and text understanding. Wishing also to educate text-mining developers, they cover basic elements that different text-mining applications include: Information extraction, topic tracking, summarization, categorization, clustering, concept linkage, information visualization, and question answering.

2.1.2 Understanding the Still Image

The visual image is a powerful channel to convey crucial information towards e-shoppers and influence their choice. In Di et al. (2014), authors investigate a well-known online marketplace, where over millions of products change hands; most are described with the help of one or more images. They present a systematic data mining and knowledge discovery approach which aims to analyze the role of images in e-commerce quantitatively and in depth. They also study interaction of image data with other selling dimensions by jointly modeling them with data from user behavior.

2.1.3 Understanding the Moving Image

In Jodoin (2013), the author presents a novel method to extract dominant motion patterns (MPs) and the main entry/exit areas from a surveillance video. The method first computes motion histograms for each pixel and then converts the histograms into orientation distribution functions. Given these functions, a novel particle meta-tracking procedure is launched which produces meta-tracks, i.e. particle trajectories. As opposed to conventional tracking, which focuses on individual moving objects, meta-tracking uses particles to follow the dominant flow of the traffic. In the last step, a novel method is used to simultaneously identify the main entry/exit areas and recover the predominant MPs. The meta-tracking procedure is a unique way to connect low-level motion features to long-range MPs. This kind of tracking is inspired by brain fiber tractography which has long been used to find dominant connections in the brain. The method is fast, simple to implement, and works both on sparse and extremely crowded scenes. It also works on highly structured scenes (highways, traffic-light corners, etc.), as well as on chaotic scenes.

2.2 Sentiment Detection Oriented Research

This section covers the research related to sentiment analysis, and is mostly oriented to social networks, and neuro-economy. Again, each research contribution is covered very briefly, only with one paragraph and one representative reference.

2.2.1 Sentiment Analysis on the Social Networks

The rising popularity of online social networks, such as Twitter, Facebook, MySpace, and LinkedIn, in recent years has sparked great interest in sentiment analysis on their data. While many methods exist for identifying sentiment in online social networks such as communication pattern mining and classification based on emoticon and parts of speech, the majority of them utilize a suboptimal batch mode learning approach when analyzing a large amount of real time data. As an alternative in Aston et al. (2014), the authors present a stream algorithm using Modified Balanced Winnow for sentiment analysis on online social networks. Tested on three real-world network datasets, the performance of these sentiment predictions is close to that of batch learning with the ability to detect important features dynamically for sentiment analysis in data streams. These top features reveal key words important to the analysis of sentiment.

2.2.2 Sentiment Analysis Using Neuro-Economy

The past decade has seen a tremendous increase in the use of neuro-physiological methods to better understand marketing phenomena among academics and practitioners. Using a unique experimental protocol to assess subjects' responses to 30-second TV ads, the authors in Venkatraman et al. (2014) capture many measures of advertising effectiveness across six commonly used methods (traditional self-reports, implicit, eye tracking, biometrics, EEG, and fMRI). These measures are shown to reliably tap into higher-level constructs commonly used in advertising research: Attention, affect, memory, and desirability. Using time-series data on sales and Gross Ratings Points for the same TV ads, the authors attempt to relate individual-level response neuro-physiological measures when participants viewed the ads in the lab to their aggregate, market-level elasticities. The experiments show that functional magnetic resonance imaging measures explain the most variance in advertising elasticities beyond the baseline traditional measures. Notably, activity in the ventral striatum is the strongest predictor of real-world, market-level response to advertising. These findings have significant implications for theory, research, and practice.

References

Aston N et al (2014) Sentiment analysis on the social networks using stream algorithms. J Data Anal Inf Process, Scientific Research 2:60–66

Di W, Sundaresan N, Piramuthu R, Bhardwaj A (2014) Is a picture really worth a thousand words?: on the role of images in e-commerce. In: WSDM '14: Proceedings of the 7th ACM international conference on Web search and data mining. ACM New York, USA, pp 633–642

Fan W et al (2006) Tapping the power of text mining. Communications of the ACM—privacy and security in highly dynamic systems, ACM New York, USA, 49(9):76–82

Jodoin PM (2013) Meta-tracking for video scene understanding. In: Proceedings on 10th IEEE international conference on advanced video and signal based surveillance (AVSS), August 2013. IEEE, Krakow, pp 1–6

Venkatraman V et al (2014) Predicting advertising success beyond traditional measures: new insights from neurophysiological methods and market response modeling. In: social science research network, internet. http://papers.ssrn.com/sol3/papers.cfm?abstract_id=2498095, 25 Oct 2014

Chapter 3
Case Studies

Abstract This chapter gives a relatively detailed overview of various successful projects involving Mind Genomics. Projects are analyzed and discussed that cover Mind Genomics based marketing activities in a number of different fields, from retail and hospital issues, all the way to education and research issues. To make it easier to comprehend for interested readers, the entire material is presented in two ways: In a visually effective form without redundancy and in the text form with appropriate redundancy.

Keywords Mind Genomics process · Marketing activities · Marketing for hospitals · Marketing for education · Marketing for retail stores · Marketing for service vendors

In this section we will be presenting a few representative case studies where Mind Genomics proved to be an excellent tool in rising customer interest. In the previous sections we described Mind Genomics to be a science that permits the provider of goods or services, or really anyone or any institution that interacts with public:

- First, to group members of that public in two, three, or maximally four different mind set (mind type) groups,
- Second, to associate a member of that public with a defined mind set group and,
- Third, to address him/her with the best possible combination of sentences and words to bring him/her closer to accepting goods, services, and other offered items.

Besides being a science considering what it contains and how it functions, Mind Genomics is really an art, an art of learning how to approach an individual, how to relate him or her to previously determined groups (mind-types), what to say to him/her, and what reactions to expect from that individual in such a process.

Why is Mind Genomics so different and so successful? Mind Genomics does not ask direct questions like "How would you like our potato-chips to taste or look" or "What do you expect when you come to your hairdresser". No. By presenting 40 or more screens, each one with three to four carefully chosen sentences that touch six main characteristics (silos) of the product/service, it goes deep into

V. Milutinovic and J. Salom, *Mind Genomics*, SpringerBriefs in Business,
DOI 10.1007/978-3-319-39733-7_3

the mind of the pollee, and makes him/her answer using feelings and not thoughts directly.

That makes the difference. The science is just there behind, to support all that and to produce the winning slogans for each one of the mind-set groups, and afterwards, to produce a few questions for typing the customer in order to place him/her into one of those mind-set groups.

Mind Genomics can be applied in almost any branch of industry (for any type of a product), any type of a service, in education, law, government, elections, and people relations—virtually everything.

We will be presenting here chosen case studies using pictures, with just one short paragraph accompanying each of them, believing that "one picture tells more than a thousand of words". First we will shortly introduce the case study, what the goals were, what, if any, variations in the approach were used in the particular case, and, finally, what the outcomes were. Then we present PowerPoint slides that visually describe the whole procedure each accompanied with a short explanation.

In the first case study the accent is placed on the procedure and in the second the accent is on the analysis of the results. Both of them are presented in greater detail for the reader to get the full picture. The remaining ones are presented only with most relevant slides.

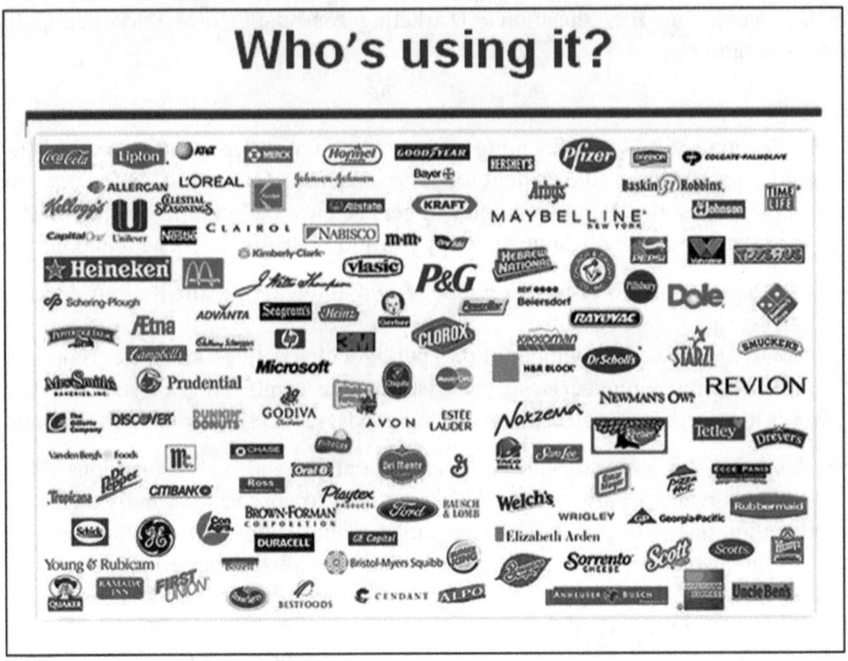

The figure above presents logos of some companies that used/are using Mind Genomics in their everyday life.

3.1 Case #1—Veterinary Hospital and Pet Care Insurance

The first case study describes how Mind Genomics was used to increase the number of users/visitors of a veterinary hospital who accept pet health care insurance that was offered. The presentation starts with three slides that nicely picture the real goals of each Mind Genomics campaign—"Imagine that you know exactly what to say to win over each person that comes to your office door". In particular, the aim was to identify how pet health care insurance can be presented to US pet owners and veterinarians in order to maximize sales conversions. It is interesting to note that the smallest increase (25 %) was obtained when an automatic interface was used (WEB), bigger increase (40 %) was achieved when a person to person interaction was used (telephone), and the largest increase obtained (50 %) was when a user visited the veterinary hospital in person. The company-wide sales increased by 42 %. The slides picture all the steps of the full Mind Genomics procedure with the ending slide presenting aforementioned increases.

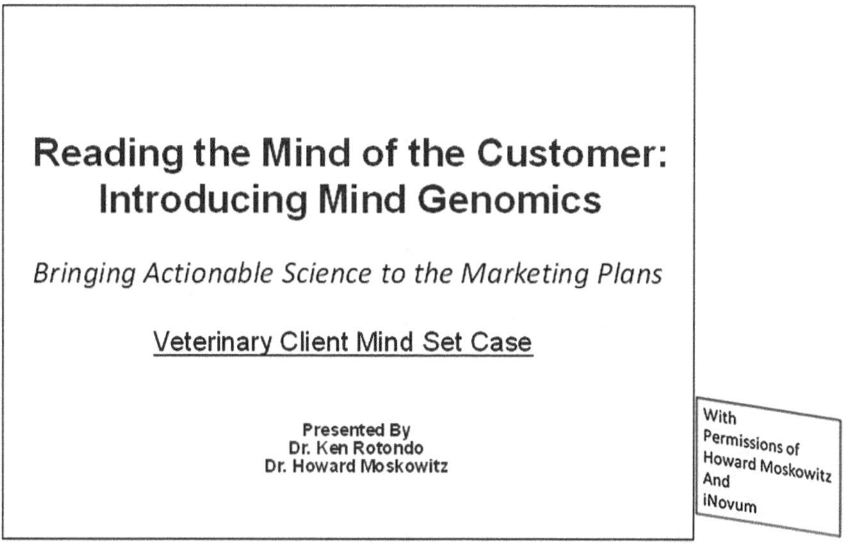

Dr. Ken Retondo and Dr. Howard Moskowitz were leading a large project oriented to the reading the mind of the customer in the specific environment related to the business of a veterinary clinic. The goal was to bring actionable science into the planning of the related marketing activities. The major results of this project are summarized in "The Veterinary Client Mind Set Case", which is presented on the pages to follow. As indicated earlier in this book, we here utilize one unorthodox presentation method; namely, for a fast reader, a PowerPoint slide represents the best use of his/her time, while the classical text is offered to slower readers concerned with details.

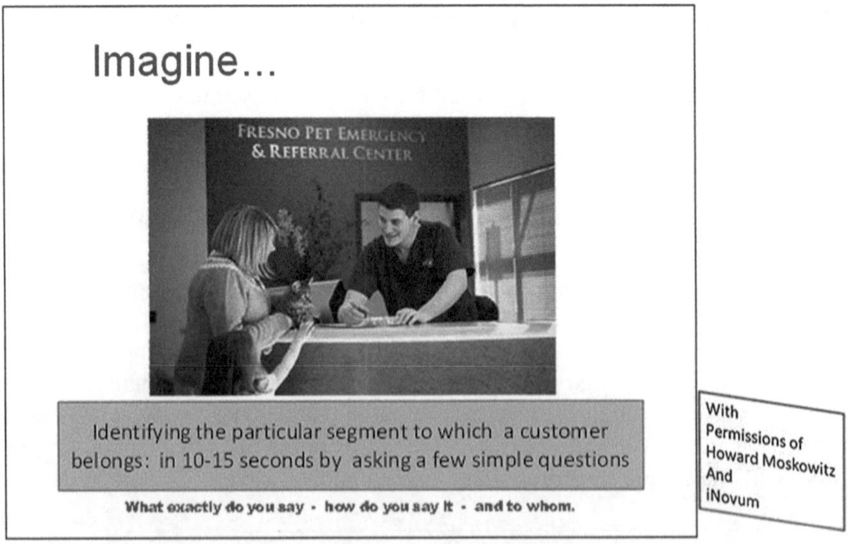

The crucial issue in each and every type of CRM (Customer Relationship Management) is how to know the customer mind type. Most of the existing methods for determining the customer mind type are penetrating into the privacy of the customer, and therefore are not convenient for the customer; actually the more useful they are to the marketing business, the less convenient they are for the customer. As indicated in the figure related to this paragraph, Mind Genomics identifies the particular segment to which a customer belongs, in only 10–15 s, by asking a few simple questions which are definitely not inconvenient to the customer to answer. As the matter of fact, the customer is highly motivated to answer those questions, due to the effects of the value exchange method utilized, as explained before.

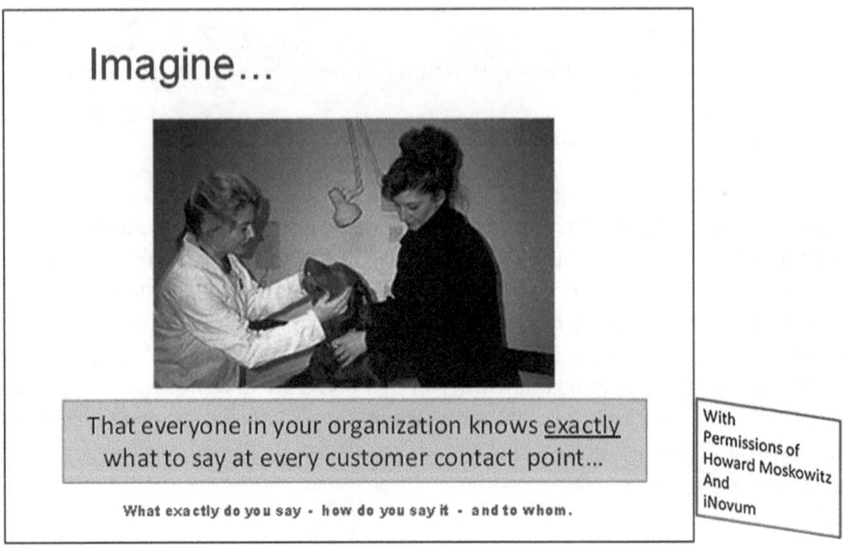

One of the results of the customers' answering the questions presented before them is that everyone in the service providing organization knows exactly what to say to each and every customer at the contact point. This is not related to the analysis and direct utilizations of the customer's answers. It is related to customer's answers in a very indirect way. The answers help the Mind Genomics system to classify the customer into one of possible customer mind types. These mind types are previously defined and characteristics of people belonging to each specific mind type are determined at the time when the mind types are set. So the questions to pose and the comments to make at contact points are stemming from the definitions of the mind type to which the customer is classified.

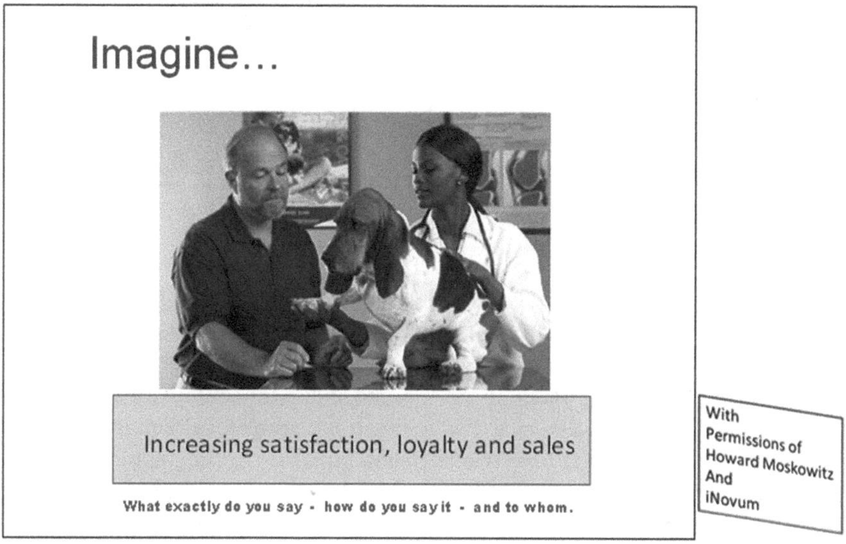

The right conversation with the customer is a key in making the customer satisfied with the services provided. Each customer, actually each customer type, has his/her own typical life goals, life priorities, and life expectancies. Being in-line with all these, increases the level of customer satisfaction, the level of customer loyalty, and the level of the sales this customer will generate, either during the ongoing visit, or during later visits in the near or far future. Specific needs for the particular type of service offered by the business may not reappear in adjacent units of time (meaning, on the daily, weekly, or the monthly bases). They may reappear a year later or years later. In such conditions, the level of awareness related to the existence and the helpfulness of a given service oriented business, will be highly correlated with the impact that the business to customer conversation had, at the first point of contact.

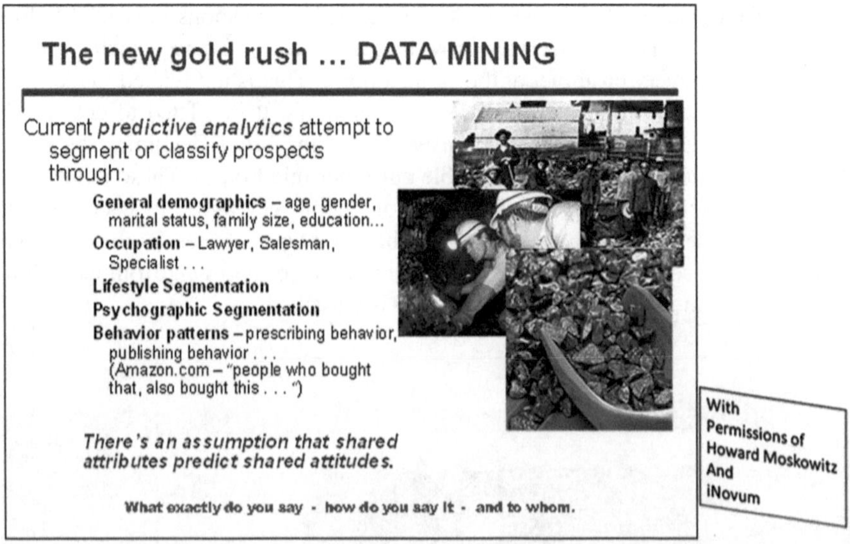

Each business has only one opportunity to make the best first impression on the customer. Therefore, the definition of possible mind types and the generation of the questioner for classifying the incoming customer into one of possible mind types have to be done very carefully, with the help of all present-day technologies related to Internet and data mining. The procedure for setting mind types used by most up-to-date predictive analytics is based on: (a) general demographic, (b) professional occupation, (c) life style segmentation, (d) psychographic segmentations, and (e) behavior patterns. Such an approach is based on one not very correct assumption that "shared attributes predict shared attitudes". Mind Genomics uses these criteria only in the process of choosing 300 pollees during its phase #1. For example, if the process is concerned with jewelry, only members of the middle and high class would be polled. In this particular case, only the people who have pets were polled. During the phase #2, Mind Genomics does not consult these criteria.

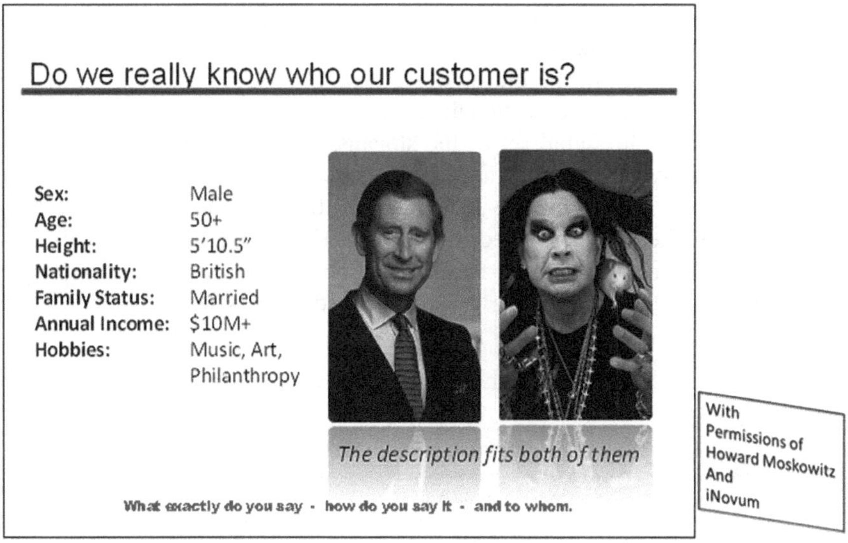

As commented for the previous figure, it is important to underline that in the case of Mind Genomics, issues related to sociology, economy, and demographics, are relevant only when Mind Genomics experts are choosing what groups of people will be polled. Simply, it is necessary to have only relevant individuals who have some experience with the topic processed taking part in the polling. Results of each polling and of the whole Mind Genomics science show that two persons of completely different social status (a prince and a musician), of a completely different economical status (both being in completely different tax brackets), and of a completely different demographic status (age, origin, education, etc.) may both love the orange color and may both be terrified of the red color. The similarities may be visible in the most subtle elements of the mind types.

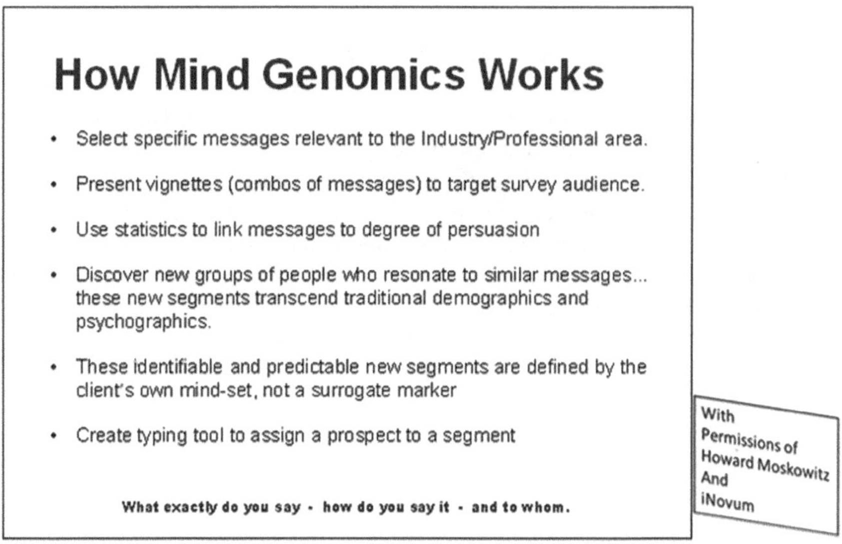

As indicated before, the Mind Genomics process implies two working phases. In the phase #1, the marketing slogans are generated. Marketing slogans are specific messages relevant to the industry/professional area. Using a specific selection process, which can be conditionally described as a random process, vignettes are generated from the initial set of the slogans. One vignette includes three or four slogans. In psychology it is known, once a person is exposed to three or four uncorrelated facts, one fills in the semantic gaps between these facts, by adding appropriate new fact based on his or her life experience. The goal is to induce an opinion in the mind of a pollee—to like it or not to like it, to be ready to pay a lot or a little. By analyzing the answers related to presented vignettes, Mind Genomics system creates customer mind types.

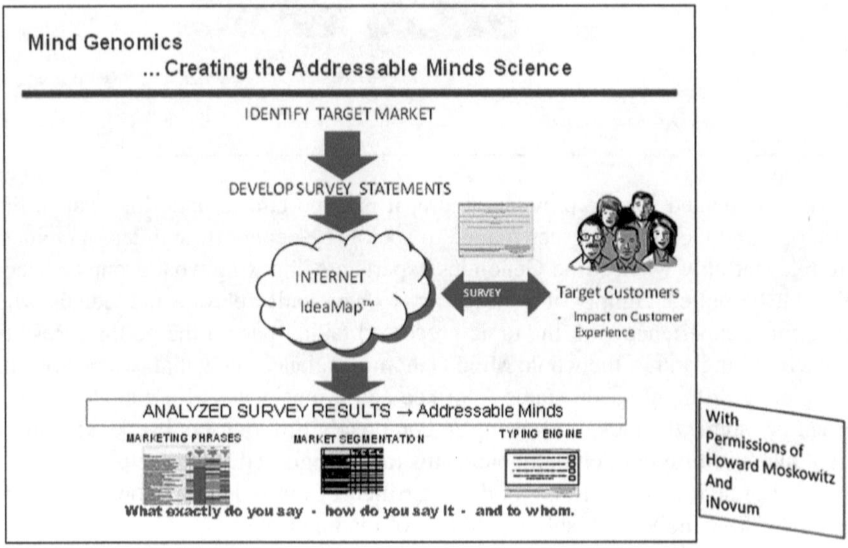

Of course, one has to first identify the target market before one can generate the marketing slogans, and to create the relevant vignettes afterwards. Using the IdeaMap software, the Mind Genomics system approaches the target customers, presents the vignettes before them (40 or 48 typically), and analyzes the results of the polling process. As said before, the resulting customer types are the end result of the phase #1. The prospects clearly falling into one of the created mind types are referred to as Addressable Minds, no matter if they were involved in the polling process or not. So far it was indicated that the polling process implies only two questions—"How do you like it?" and "How much are you willing to pay for it?". In principal, a third question could be added, too: "What is your opinion about it?".

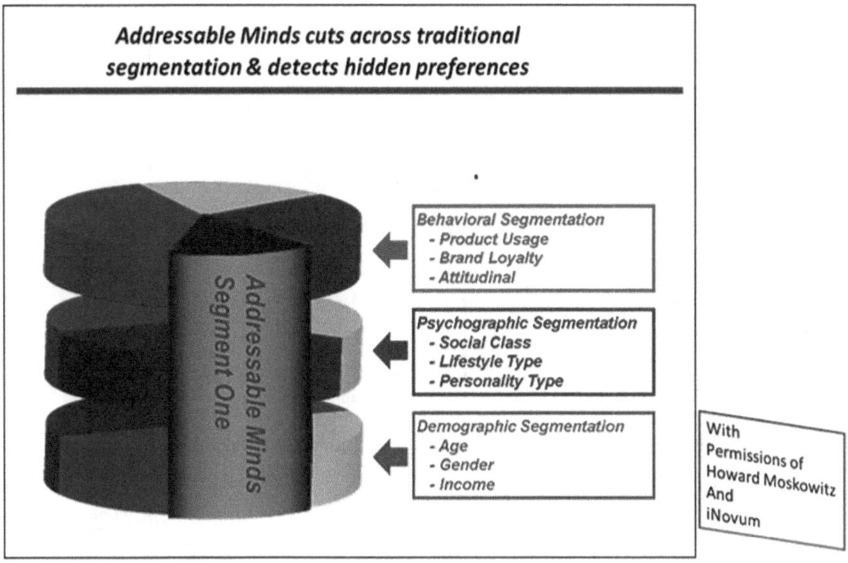

During the polling activity, which is a part of Phase #1, one has to determine what the right group of people to use is. Selection implies utilization of three types of segmentation: (a) behavioral segmentation (product usage, brand loyalty, and related attitudes), (b) psychological segmentation (social class. lifestyle habits, and personality types), and (c) demographic segmentation (age, gender, and income), but as mentioned before, if the subject of the process by definition excludes some group because its members by definition are not possible under of the product or service, such groups are excluded from the polling. Proper selection along the mentioned three segmentation domains generates proper Addressable Minds groups, but not defined within the boundaries of any specific segmentation. The Mind Genomics process cuts across those boundaries and the mind types are defined only by using the polling answers and not by taking into account who those answers come from.

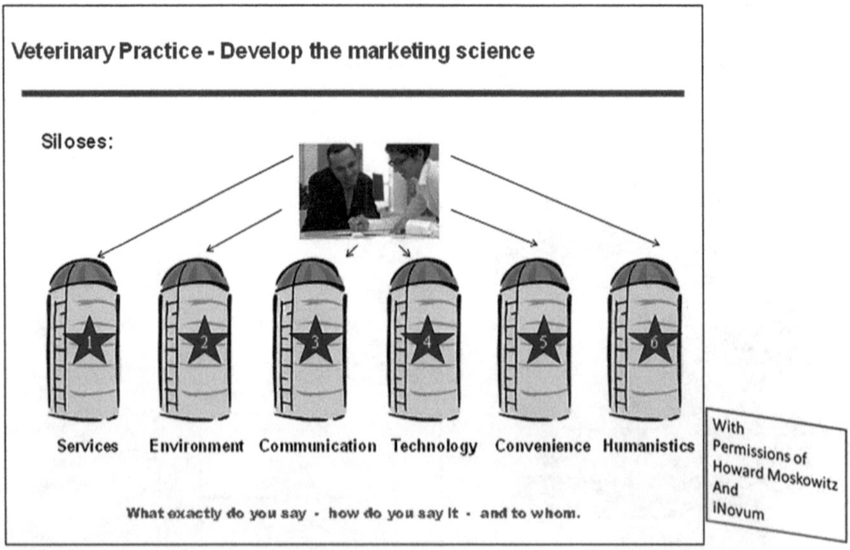

It is extremely important that the marketing slogans are properly defined. As indicated before, a good number of marketing slogans to be defined is 36. Such a relatively large number of marketing slogans is impossible to generate properly unless proper guidelines are set first. Existing guidelines maintain that one first has to generate six silos. Selection of silos has to be done in cooperation with a field expert, who has a proper understanding of all essential aspects of the business for which the slogans are to be generated. For the case of the veterinary business being the subject of our present overview, the selected silos, as presented in the figure accompanying this paragraph are: (a) services, (b) environment, (c) communication, (d) technology, (e) convenience, and (f) humanistics.

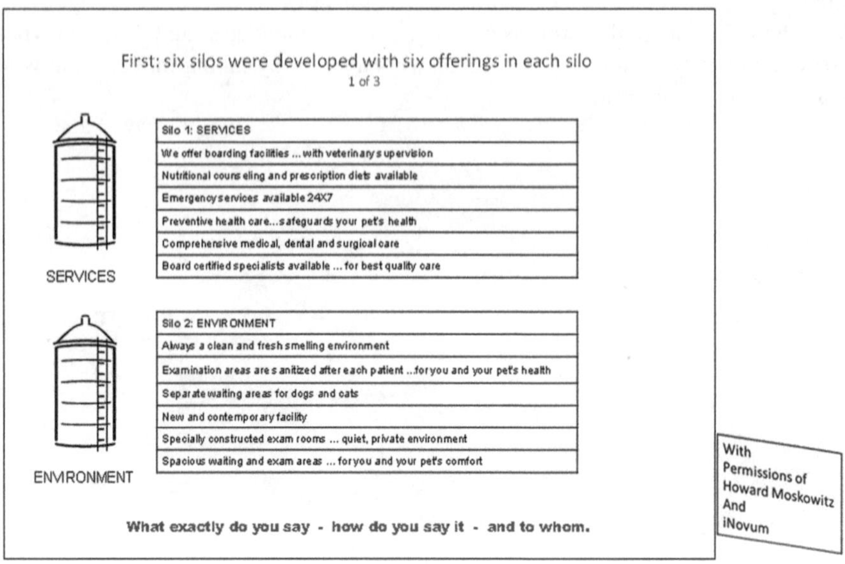

Once the silos are created, for each and every silo some number of slogans has to be created. One good number is six slogans per a silo. In general, there is a freedom in creating the wording of the slogans; however, some researchers suggest that each slogan should include three small semantic entities, one referring to the resident silo, and two other ones referring to two of the remaining five silos. The resident silo is the one that includes all slogans in which the first semantic entity is related to it. In the figures accompanying this paragraph, two silos and their corresponding slogans are presented. The examples from slogans in the silo #1—SERVICES are: “Emergency services available 24 × 7”, “We offer boarding facilities … with veterinary supervision”, and “Comprehensive medical, dental and surgical care”. The examples from slogans in the silo #2—ENVIRONMENT are: “Always a clean and fresh smelling environment”, “Examination areas are sanitized after each patient … for your pet’s health”, and “Separate waiting areas for dogs and cats”.

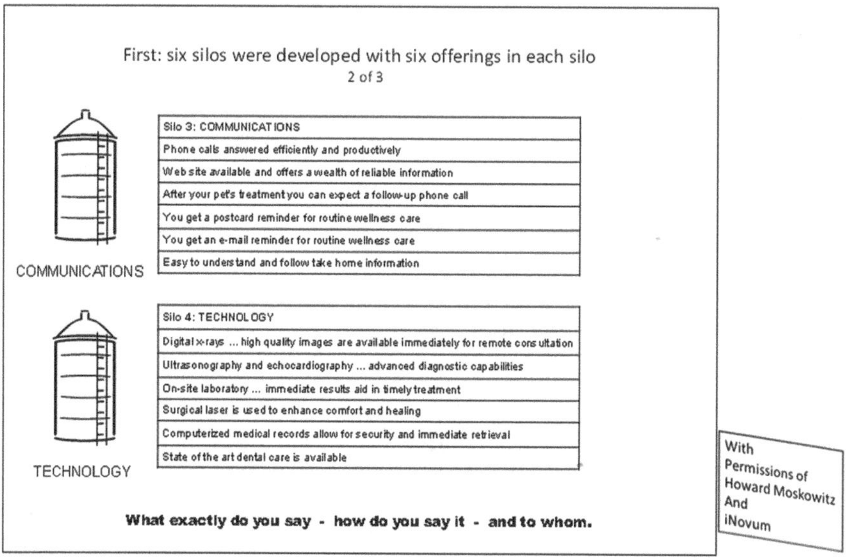

The silo #3—COMMUNICATIONS includes the following six slogans: (a) Phone calls answered efficiently and productively, (b) Web site available and offers a wealth of reliable information, (c) After your pet’s treatment you can expect a follow-up phone call, (d) You get a postcard reminder for routine wellness care, (e) You get an e-mail reminder for routine wellness care, and (f) Easy to understand and follow take home information. The silo #4—TECHNOLOGY includes the following six slogans: (a) Digital x-rays … high quality images are available immediately for remote consultation, (b) Ultrasonography and echocardiography … advanced diagnostic capabilities, (c) On-site laboratory … immediate results aid in timely treatment, (d) Surgical laser is used to enhance comfort and healing, (e) Computerized medical records allow for security and immediate retrieval, and (f) State of the art dental care is available.

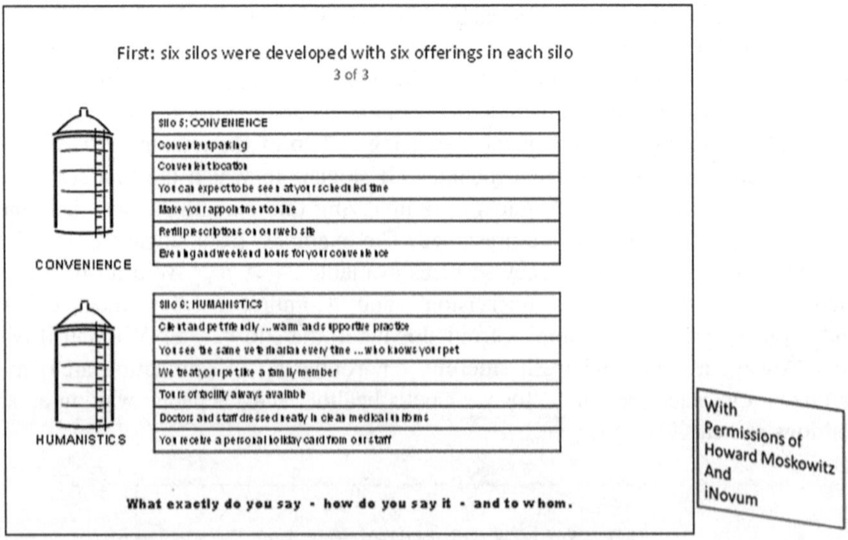

The silo #5—CONVENIENCE includes the following six slogans: (a) Convenient parking, (b) Convenient location, (c) You can expect to be seen at your scheduled time, (d) Make your appointment online, (e) Refill prescriptions on our web site, and (f) Evening and weekend hours for your convenience. The silo #6—HUMANISTICS includes the following six slogans: (a) Client and pet friendly … warm and supportive practice, (b) You see the same veterinarian every time … who knows your pet, (c) We treat your pet like a family member, (d) Tours of the facility always available, (e) Doctors and staff dressed neatly in clean medical uniforms, and (f) You receive a personal holiday card from our staff.

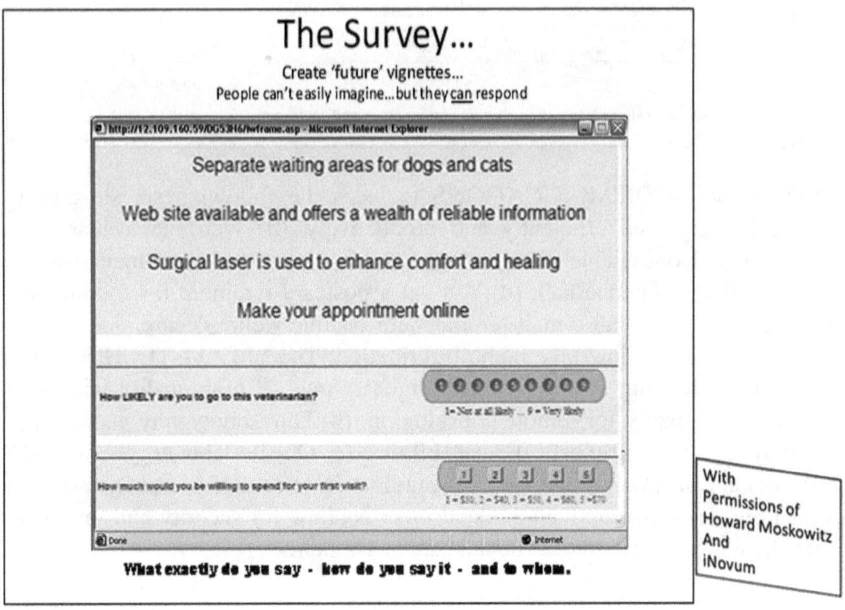

The figure accompanying this paragraph shows one possible layout of the screen used to present a vignette and to ask related questions. First, one should answer how these slogans attract him/her to come to this vet hospital, by rating this "likelyness (probability) of a visit" from 1 ("Not at all likely") to 9 ("Very likely"). Please note that the possible answers to the question how much one is willing to spend for their first visit is not quantitative (like: "I am willing to pay $1, or $10, or $100, or $1000, or $10,000). The possible answers are qualitative (like "I am not ready to pay anything" graded 1, "I am ready to pay something"—graded from 2 to 4, and "I am ready to pay a lot"—graded 5). The number of screens of this sort is equal to the number of vignettes which is, as indicated before, 40 or 48 in most cases.

The Results… what pet owners want from a personal preference perspective –

Veterinary Team leads client communications with the green & avoid the red

	TOTAL MARKET	Seg 1 WANTS PROMPT & WARM PERSONAL SERVICE	Seg 2 SEEKS TECHNICALLY COMPETENT PROFESSIONALS	Seg 3 PLEASE TREAT MY PET JUST LIKE A PERSON
BASE SIZE:	(109)	(36)	(36)	(37)
PROPENSITY TO BE A CUSTOMER:	**66**	**85**	**49**	**65**
Emergency services available 24X7	11	-4	17	20
You see the same veterinarian every time ... who knows your pet	10	6	19	6
Board certified specialists available ... for best quality care	9	-4	17	14
We treat your pet like a family member	7	2	11	8
Evening and weekend hours for your convenience	6	4	0	15
Client and pet friendly ... warm and supportive practice	6	6	8	4
Web site available and offers a wealth of reliable information	6	8	9	1
On site laboratory . Immediate results aid in timely treatment	6	-1	19	0
Digital x-rays ... high quality images are available immediately for remote consultation	6	-4	25	-1
Convenient location	-1	2	-2	-3
Tours of facility always available	-1	-4	4	-4
Doctors and staff dressed neatly in clean medical uniforms	-1	-1	4	-6
Make your appointment online	-2	-5	-9	7
Spacious waiting and exam areas ... for you and your pet's comfort	-2	-3	3	-7
Separate waiting areas for dogs and cats	-3	5	2	-15
You receive a personal holiday card from our staff	-4	-4	9	-16

Advocacy Messages ≥ 6 · Detractor Messages ≤ -1 · Neutral Messages

Positive communications are green. Avoid all red communications. Non-color communications are neutral.

What exactly do you say - how do you say it - and to whom.

With Permissions of Howard Moskowitz And iNovum

The final activity in the phase #1 is the analysis that creates mind types and defines their respected characteristics. It is recommended that the number of generated mind types is only three, in most cases. In principle, it can be as high as 9 or 27; however, in those two cases, the probability of misplacing an incoming prospect into a wrong mind type may become prohibitively high. Also, the lower the number of mind types, the higher is the robustness of the system in relation to answers which are not given honestly, or are given in a rush, without much thinking about the essence of the question posed during the polling process. The figure related to this paragraph has four columns with numbers. The first column (Total market) indicates how many pollees of the total polled population voted in favor of the slogan (green cell) and how many disliked it, or were indifferent toward it (red cell).

The Results

... Segment 1 - WANTS PROMPT & WARM PERSONAL SERVICE

	TOTAL MARKET	Seg 1 WANTS PROMPT & WARM PERSONAL SERVICE	Seg 2 SEEKS TECHNICALLY COMPETENT PROFESSIONALS	Seg 3 PLEASE TREAT MY PET JUST LIKE A PERSON
BASE SIZE:	(109)	(36)	(36)	(37)
PROPENSITY TO BE A CUSTOMER:	66	85	49	65
Web site available and offers a wealth of reliable information	6	8	9	1
Phone calls answered efficiently and productively	0	7	-10	3
You see the same veterinarian every time ... who knows your pet	10	6	19	6
Client and pet friendly ... warm and supportive practice	6	6	8	4
Specially constructed exam rooms ... quiet, private environment	0	6	3	-10
On site laboratory . Immediate results aid in timely treatment	6	-1	19	0
You get a postcard reminder for routine wellness care	0	-1	-3	2
Doctors and staff dressed neatly in clean medical uniforms	-1	-1	4	-6
Refill prescriptions on our web site	0	-2	-5	7
Ultrasonography and echocardiography ... advanced diagnostic capabilities	5	-3	19	-1
State of the art dental care is available	4	-3	16	-2
Easy to understand and follow take home information	2	-3	6	1
Spacious waiting and exam areas ... for you and your pet's comfort	-2	-3	3	-7
Emergency services available 24X7	11	-4	17	20
Board certified specialists available ... for best quality care	9	-4	17	14
Digital x-rays ... high quality images are available immediately for remote consultation	6	-4	25	-1
We offer boarding facilities ... with veterinary supervision	5	-4	11	9
Tours of facility always available	-1	-4	4	-4
You receive a personal holiday card from our staff	-4	-4	9	-16
Convenient parking	1	-5	4	4
Preventive health care safeguards your pet's health	1	-5	5	3
Surgical laser is used to enhance comfort and healing	1	-5	15	-5
Make your appointment online	-2	-5	-9	7
Computerized medical records allow for security and immediate retrieval	2	-6	12	1
Comprehensive medical, dental and surgical care	5	-7	8	13
Nutritional counseling and prescription diets available	4	-9	13	8

What exactly do you say - how do you say it - and to whom.

With Permissions of Howard Moskowitz And iNovum

The figure related to this paragraph shows how one establishes a mind type and how one decides what major characteristics of such a mind type are. The impact of the number of pollees picking the same answers, and the scale at which they answered is crucial for determining the characteristics of the particular mind type. Sophisticated math apparatus based on linear regression is a tool that selects groups of pollees that answered similarly. This table is sorted using values in column labeled Segment 1. It presents what slogans are mostly appreciated by the first mind set group named “Wants prompt & warm personal service”, and which slogans are mostly ignored or disliked. Most favorable feeling pollees grouped in mind type #1 had for the following slogans: “Web site available and offers a wealth of reliable information”, “Phone calls answered efficiently and productively”, and “You see the same veterinarian every time … who knows your pet”. The slogan that was mostly disliked was “Nutritional counseling and prescription diets available”.

The Results

... Segment 2 – SEEKS TECHNICALLY COMPETENT PROFESSIONALS

	TOTAL MARKET	Seg 1 WANTS PROMPT & WARM PERSONAL SERVICE	Seg 2 SEEKS TECHNICALLY COMPETENT PROFESSIONALS	Seg 3 PLEASE TREAT MY PET JUST LIKE A PERSON
BASE SIZE:	(109)	(36)	(36)	(37)
PROPENSITY TO BE A CUSTOMER:	**66**	**85**	**49**	**65**
Digital x-rays ... high quality images are available immediately for remote consultation	6	-4	25	-1
You see the same veterinarian every time ... who knows your pet	10	6	19	6
On site laboratory . immediate results aid in timely treatment	6	-1	19	0
Ultrasonography and echocardiography ... advanced diagnostic capabilities	5	-3	19	-1
Emergency services available 24X7	11	-4	17	20
Board certified specialists available ... for best quality care	9	-4	17	14
State of the art dental care is available	4	-3	16	-2
Examination areas are sanitized after each patient ...for you and your pet's health	5	4	15	-5
Surgical laser is used to enhance comfort and healing	1	-5	15	-5
Nutritional counseling and prescription diets available	4	-9	13	8
Computerized medical records allow for security and immediate retrieval	2	-6	12	1
We treat your pet like a family member	7	2	11	8
We offer boarding facilities ... with veterinary supervision	5	-4	11	9
Web site available and offers a wealth of reliable information	6	8	9	1
You receive a personal holiday card from our staff	-4	-4	9	-16
Client and pet friendly ... warm and supportive practice	6	6	8	4
Comprehensive medical, dental and surgical care	5	-7	8	13
Easy to understand and follow take home information	2	-3	8	1
After your pet's treatment you can expect a follow-up phone call	3	3	-1	6
Convenient location	-1	2	-2	-3
You get a postcard reminder for routine wellness care	0	-1	-3	2
You can expect to be seen at your scheduled time	3	4	-5	12
Refill prescriptions on our web site	0	-2	-5	7
You get an e-mail reminder for routine wellness care	0	3	-7	3
Make your appointment online	-2	-5	-9	7
Phone calls answered efficiently and productively	0	7	-10	3

What exactly do you say - how do you say it - and to whom.

With Permissions of Howard Moskowitz And iNovum

The table in the figure related to this paragraph is sorted using values in column labeled Seg. 2. It presents how slogans are valued by the second mind set group named "Seeks technically competent professionals". The mostly appreciated slogans were: "Digital x-rays … high quality images are available immediately for remote consultation", "You see the same veterinarian every time … who knows your pet", "On site laboratory … immediate results aid in timely treatment", "Ultrasonography and echocardiography … advanced diagnostic capabilities", "Emergency services available 24 × 7", "Board certified specialists available … for best quality care", "State of the art dental care is available", "Examination areas are sanitized after each patient …for you and your pet's health", and "Surgical laser is used to enhance comfort and healing". The slogans that were mostly ignored or rejected were: "Phone calls answered efficiently and productively", "Make your appointment online", "You get an e-mail reminder for routine wellness care", "Refill prescriptions on our web site", and "You can expect to be seen at your scheduled time".

The Results

. . . Segment 3 – PLEASE TREAT MY PET JUST LIKE A PERSON

	TOTAL MARKET	Seg 1 WANTS PROMPT & WARM PERSONAL SERVICE	Seg 2 SEEKS TECHNICALLY COMPETENT PROFESSIONALS	Seg3 PLEASE TREAT MY PET JUST LIKE A PERSON
BASE SIZE:	(109)	(36)	(36)	(37)
PROPENSITY TO BE A CUSTOMER:	**66**	**85**	**49**	**65**
Emergency services available 24X7	11	-4	17	20
Evening and weekend hours for your convenience	6	4	0	15
Board certified specialists available ... for best quality care	9	-4	17	14
Comprehensive medical, dental and surgical care	5	-7	8	13
You can expect to be seen at your scheduled time	3	4	-5	12
We offer boarding facilities ... with veterinary supervision	5	-4	11	9
We treat your pet like a family member	7	2	11	8
Nutritional counseling and prescription diets available	4	-9	13	8
Refill prescriptions on our web site	0	-2	-5	7
Make your appointment online	-2	-5	-9	7
You see the same veterinarian every time ... who knows your pet	10	6	19	6
After your pet's treatment you can expect a follow-up phone call	3	3	-1	6
Digital x-rays ... high quality images are available immediately for remote consultation	6	-4	25	-1
Ultrasonography and echocardiography ... advanced diagnostic capabilities	5	-3	19	-1
State of the art dental care is available	4	-3	16	-2
Always a clean and fresh smelling environment	1	3	2	-2
New and contemporary facility	0	0	3	-2
Convenient location	-1	2	-2	-3
Tours of facility always available	-1	-4	4	-4
Examination areas are sanitized after each patient ...for you and your pet's health	5	4	15	-5
Surgical laser is used to enhance comfort and healing	1	-5	15	-5
Doctors and staff dressed neatly in clean medical uniforms	-1	-1	4	-6
Spacious waiting and exam areas ... for you and your pet's comfort	-2	-3	3	-7
Specially constructed exam rooms ... quiet, private environment	0	6	3	-10
Separate waiting areas for dogs and cats	-3	5	2	-15
You receive a personal holiday card from our staff	-4	-4	9	-16

What exactly do you say - how do you say it - and to whom.

With Permissions of Howard Moskowitz And INovum

The table in the figure related to this paragraph is sorted using values in column labeled Seg. 3. It presents how slogans are valued by the third mind set group named "Please treat my pet just like a person". The mostly appreciated slogans were: "Emergency services available 24 × 7". "Evening and weekend hours for your convenience", "Board certified specialists available … for best quality care", "Comprehensive medical, dental and surgical care", "You can expect to be seen at your scheduled time", "We offer boarding facilities … with veterinary supervision", "We treat your pet like a family member", and "Nutritional counseling and prescription diets available". The slogans that were mostly ignored or rejected were: "You receive a personal holiday card from our staff", "Separate waiting areas for dogs and cats", and "Specially constructed exam rooms … quiet, private environment".

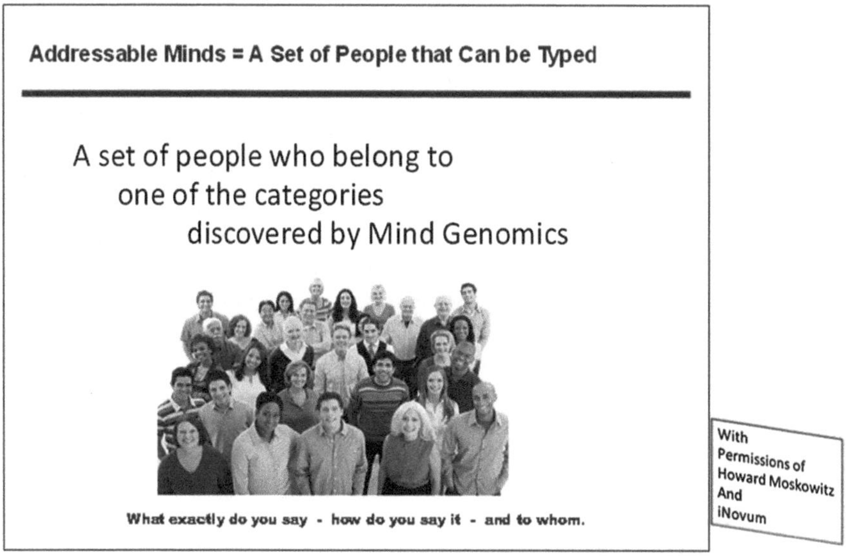

The marketing slogans addressing the three mind types should be chosen from the best valued answers during the polling process. For example, for the mind type #1, they should be "Web site available and offers a wealth of reliable information", "Phone calls answered efficiently and productively", and "You see the same veterinarian every time … who knows your pet". For the mind type #2, they should be, for example, "Digital x-rays … high quality images are available immediately for remote consultation", "You see the same veterinarian every time … who knows your pet", and "On site laboratory … immediate results aid in timely treatment". Finally, for the mind type #3, the suggested slogans should be "Emergency services available 24 × 7", "Evening and weekend hours for your convenience", and "Board certified specialists available … for best quality care".

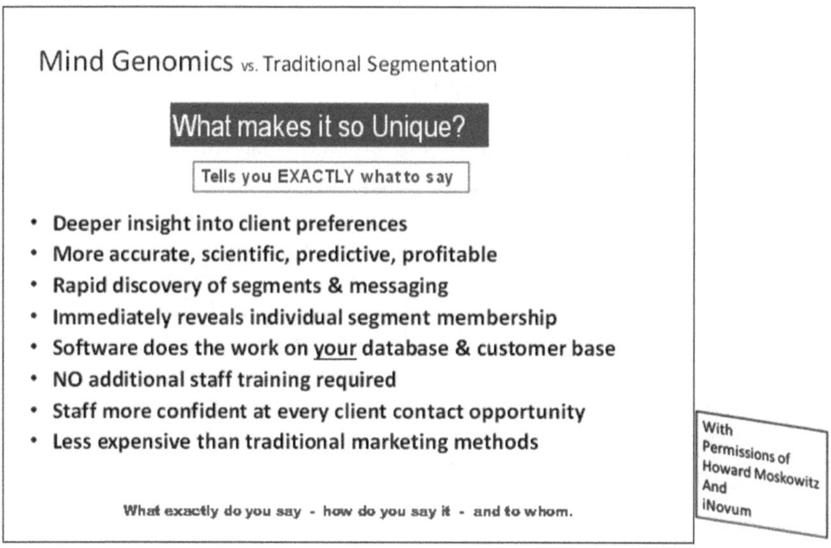

In essence, what Mind Genomics does in its phase #1 is creation of baskets that present the existing mind types on the market. What it does in its phase #2 is putting the incoming prospects into one of the created baskets; and not only that—it also provides a deeper insight into client's preferences, articulated through questions and comments to be imposed onto the customer. Those suggested questions and comments are more accurate, scientific, predictive, and profitable. They provide rapid discovery of segments and messaging immediately revealing individual segment membership. The Mind genomics software does the work on both company's own database and customer base, requires no additional staff training, and makes the staff more confident at every contact opportunity with the client. Also, using Mind Genomics as a marketing tool is much more cost effective than using traditional marketing methods.

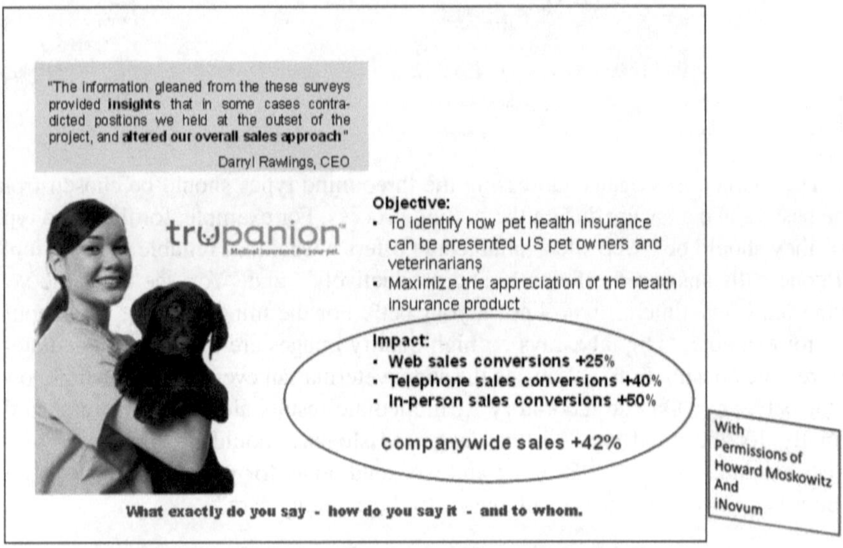

In real life, an original idea is worth its implementation effort only if it brings benefits. In marketing, "benefits" means "profits". What is also important is the impact of a long term aggregation of benefits. In the case of MindGenomics, the impacts are best understood through the percentages of keystone improvements. In the case of veterinary hospital this concretely meant: (a) web sales conversions increased by 25 % (percentage of the customers who came to the hospital and, then, decided to take the pet insurance), (b) telephone sales conversions increased by 40 %, and (c) in-person sales conversions increased by 50 %. On the company level, overall, this meant that the sales were increased by 42 %. An interesting study would be to determine the correlation between increases in sales and increases in profits. With a 42 % increase in sales, one could afford to sell at a larger profit per item.

3.2 Case #2—Louisiana Coast Spill

The Deepwater Horizon oil spill in the Gulf of Mexico began on April 20, 2010 at the BP-owned Transocean-operated Macondo Prospect. Following the explosion and sinking of the Deepwater Horizon oil rig, a sea-floor oil gusher flowed for 87 days, until it was capped on July 15, 2010. Between May and July 2010, the spill waters near the Louisiana Coast contained 40 times more Polycyclic aromatic hydrocarbons than before the spill. As a result, the whole Louisiana Coast fishing industry collapsed. Even when the sea coast was cleaned and clear, it was extremely difficult to win back the customers. The role of the following Mind Genomics campaign was to find winning slogans that could persuade American public to start again buying sea food originating from Louisiana Coast. This presentation also analyzes the results by gender, age groups, and ethnic groups. The Mind Genomics campaign was successful in bringing sales to the levels before the spill and even higher than that.

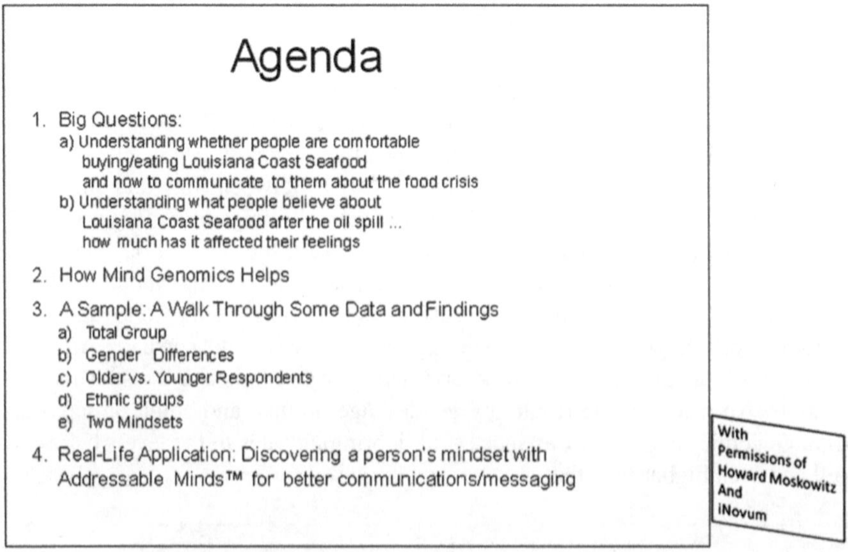

A proper plan has to be established at the beginning of each and every Mind Genomics process. It implies four basic agenda items: (a) Finding what the main questions aimed at better understanding of the overall environment are; (b) Analyzing how Mind Genomics could help in the given environment, under the given circumstances; (c) Developing a sample, which sheds more light on specific data findings; and (d) Developing a real life Addressable Minds™ application to discover a person's mindset for better communications/messaging. The above described four agenda items may vary from case to case. It is most important to follow the logic of the problem, of the environment, and of the circumstances, rather than to follow blindly some predefined algorithmic path of Mind Genomics. In other words, one has to be flexible to get the best performance gains.

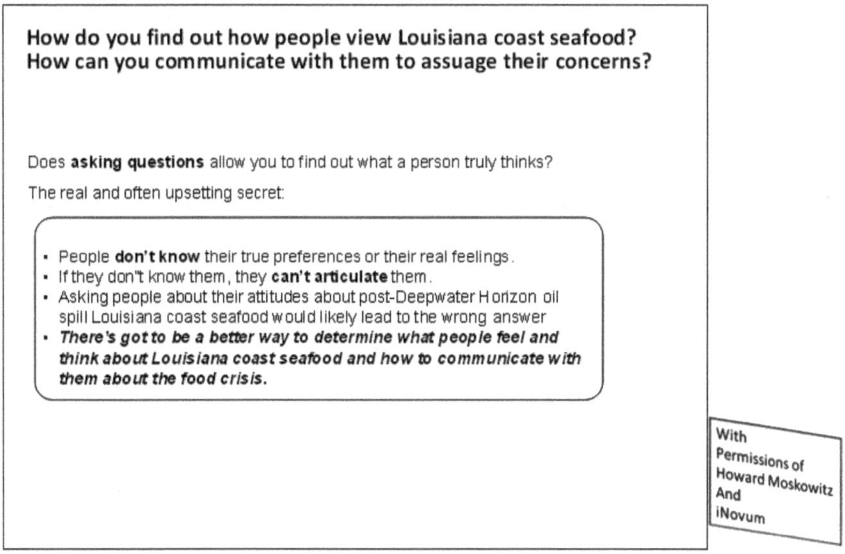

The most important issue of all is to find out how people view the problem for which mind Genomics tries to help. In this particular case, the issue was to discover how people view Louisiana coast sea food. The next most important issue is how to communicate with the people in order to learn about their concerns. In this case the issue was to figure out what specific concerns may drive the customers away from purchasing the Louisiana sea food products. This was not an easy task. As indicated in the figure that accompanies this paragraph, sometimes people don't know about their real preferences or feelings. At other times they know, but they can't articulate them. If they know and can articulate, they may misunderstand your question, which naturally leads to the wrong answer. The main point is how to avoid all the above mentioned "Scilas and Haribdas".

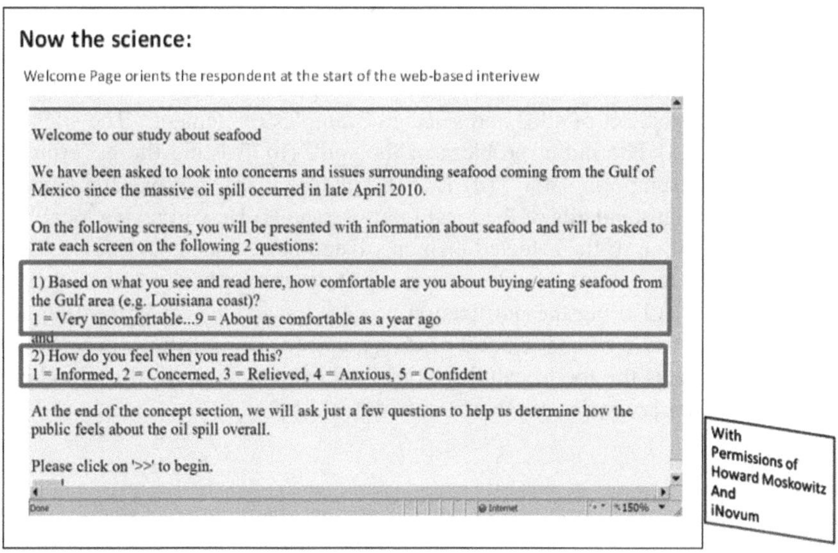

As explained before, a proper polling process has to be defined. The offered answers refer to a gradual change of preferences and/or feelings, which ensures that the probability of misunderstanding drops down considerably. Proper scaling includes terms like uncomfortable or very uncomfortable. Also terms like informed, concerned, relieved, anxious, and confident. With such a gradual progress through issues, Mind Genomics helps the pollees generate proper expectations and feelings, even in cases when they have no preferences and feelings. In cases when they do have preferences and feelings, but they cannot articulate them, the above described approach helps them articulate them. And, finally, misunderstanding becomes less likely to happen.

The Demo Study: The Structure

Material for the test ...messages about Louisiana coast seafood.
Generate 6 subject categories or silos ... each silo has 6 ideas.
Below ...6 silos ... one element from each silo.

Silos:	Elements:
1. Problem	*The oil spill is not as big of an ecological issue as it's being made out to be*
2. Frightened	*You can trust the government to protectthe food supply chain fromcontamination*
3. Seafood consumption safety	*Shrimp and crab are abundant and safe to eat*
4. Nature	*Natural occurring bacteria will act to remove oil from the Gulf ecosystem*
5. Louisiana food industry has high standards	*Louisiana's food industry is on top of seafood production standards*
6. Reviving Louisiana seafood consumption	*Buying seafood from Louisiana will help revive the seafood industry*

With Permissions of Howard Moskowitz And iNovum

Creating silos and elements (possible marketing slogans) is an activity that requires both experience and innovation. The figure related to this paragraph reveals one possible set of silos and selected examples of elements. The silos utilized here are: (a) The initial problem of the spill, (b) Trusting the government, (c) Seafood consumption safety, (d) How the nature helps improve the environment, (e) The high standards of the local industry, and (f) How to revive local food industry. In addition to the selected elements listed here, one should device other potentially effective elements, like these: (a) Our sea products are the cheapest on the market, (b) During the purification process we add cancer-preventing substances into the tissue of our seafood products; and (c) Our purification process is superior and makes the food healthier, compared to the uncorrupted seafood processed by inferior processes of other countries.

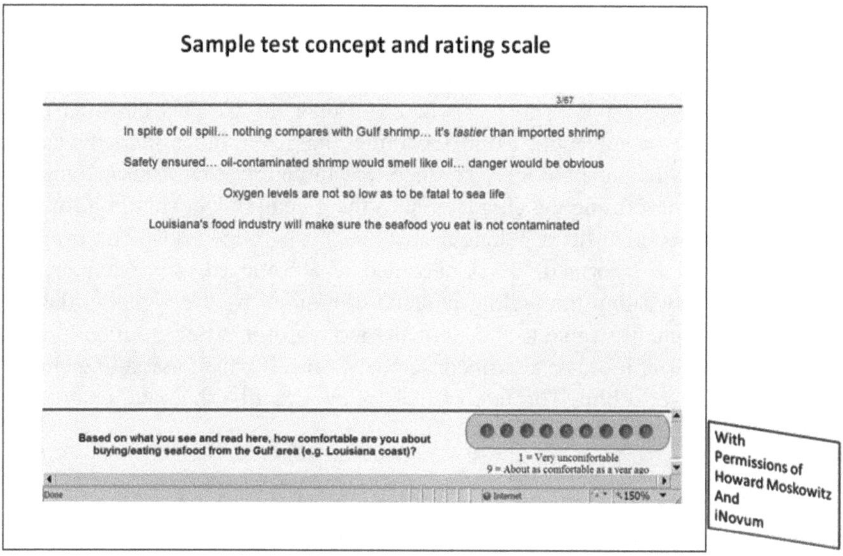

The accompanying figure shows an example of a vignette. As the reader remembers, the four elements of the vignette can be completely uncorrelated. Even if totally uncorrelated, the reader, based on the experience (s)he has, will generate a scenario which is: "like or doesn't like", "more value or less value". The purpose of these elements, correlated or uncorrelated, is to create a conscious or an unconscious picture in the mind of a pollee that can induce a clear level of appreciation or resentment toward the topic described; the level that can be expressed in numbers (from 1—"I feel very uncomfortable", to 9—"I feel as comfortable as a year ago"). As it was stated before, the pollee is not answering any direct questions, (s) he simply grades what feeling a group of sentences induces in his/her mind.

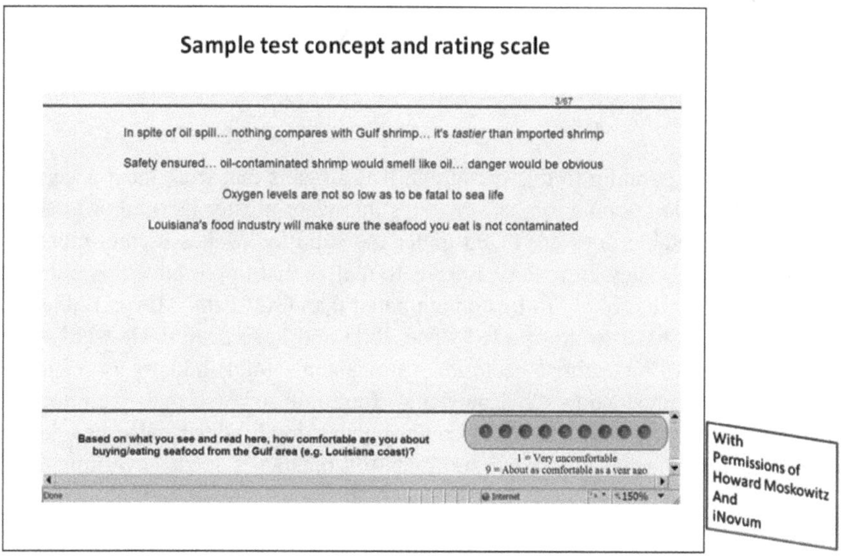

The previous vignette was accompanied with the question "Based on what you see here, how comfortable you are about buying/eating seafood from the Gulf area (e.g. Louisiana coast)?". This vignette is the same one, but accompanied with another question: "How do you feel when you read this?". This question really holds the main advantage the Mind Genomics has over other market analysis methodologies. Other technologies produce questionnaires that ask customers to answer question describing the characteristics the given product should have. This vignette asks the pollee (as a potential customer) how (s)he feels. The available answers were: 1 = Informed, 2 = Concerned, 3 = Relieved, 4 = Anxious, and 5 = Confident. By rating the feeling induced in him/her by the vignette, the pollee gives a very precise value to the elements written in it. After grading 40 or 48 vignettes, the pollee, more or less unconsciously, reveals which marketing slogans might be attractive to him. The figures that follow describe the data, results, and insights based on those data and results.

What makes you feel like you felt last year – total panel

About one in four feel the same way as a year ago ... what an effect!
Not much drives them to feel like last year.
Best option... talk about imported fish (not a safe alternative).

	Total Sample ... Do you feel the same as last year?	
	Additive constant (baseline)	24
	Makes us feel more like we felt last year	
C5	Public needs to realize that imported fish is not well-regulated and MORE likely to be contaminated than Gulf fish	5
A4	Over 12,000 oil-related incidents have been reported since 2005 and have been dealt with	4
D6	The Gulf has natural "oil eating" bacteria	4
E3	Louisiana's food industry is on top of seafood production standards	4
E4	Louisiana's food industry will make sure the seafood you eat is not contaminated	4
E6	Louisiana's seafood industry also gets its fish from areas not affected by the oil spill	3
E5	Louisiana's local government will not allow contaminated seafood tp hit the market	3
	Makes us feel different from how we felt last year	
A2	Louisiana's seafood industry employs more than 27,000 people ... why should they be punished	-2
F2	Americans' appetite for seafood won't be changed overnight	-3
A1	The oil spill is not as big of an ecological issue as it's being made out to be	-3
B2	Some people don't care where fish comes from .... regardless of the spill	-5

With Permissions of Howard Moskowitz And iNovum

The figure responding to this paragraph lists slogans that were most acceptable to the whole polled population and the ones that were mostly ignored or unacceptable. The acceptable ones are listed under the subtitle "Makes us feel more like we felt last year". They are: "Public needs to realize that imported fish is not well-regulated and MORE likely to be contaminated than Gulf fish", "Over 12,000 oil-related incidents have been reported since 2005 and have been dealt with", "The Gulf has natural "oil eating" bacteria", "Louisiana's food industry is on top of seafood production standards", "Louisiana's food industry will make sure the seafood you eat is not contaminated", "Louisiana's seafood industry also gets its fish from areas not affected by the oil spill", and "Louisiana's local government will not allow contaminated seafood to hit the market. The unattractive ones are listed under the subtitle “Makes us feel different from how we felt last year”. They are: “Louisiana’s seafood industry employs more than 27,000 people … why should they be punished”, “Americans’ appetite for seafood won’t be changed overnight”, “The oil spill is not as big of an ecological issue as it’s being made out to be”, and “Some people don’t care where fish comes from … regardless of the spill.”

What elements drive each dominant feeling: total panel

Very few elements drive strong emotional reactions
Does no one feel anything ... Or is it different mind-sets.

Emotions - Total panel	Anx	Cnd	Rel	Cnf
Anxiety drivers				
The oil spill is not as big of an ecological issue as it's being made out to be	6	3	2	1
Concerned drivers				
Louisiana's food industry is on top of seafood production standards	4	10	7	4
Relieved drivers				
Louisiana's food industry adopts the highest standards	3	7	9	3
Confident drivers				
Louisiana's food industry will make sure the seafood you eat is not contaminated	2	4	6	6

With Permissions of Howard Moskowitz And iNovum

As it was mentioned before, the second question on each vignette was “How do you feel when you read this?”. The available answers were: 1 = Informed, 2 = Concerned, 3 = Relieved, 4 = Anxious, and 5 = Confident. It is interesting to analyze what slogans were mostly chosen for each of the emotions (except for the indifferent one: “1 = Informed”). As depicted in the table in the figure related to this paragraph (a) the slogan used on vignettes that mostly caused Anxiousness was “The oil spill is not as big of an ecological issue as it’s being made out to be”; (b) the slogan that mostly caused Concern was “Louisiana’s food industry is on top of seafood production standards”; (c) the slogan that mostly caused Relief was “Louisiana’s food industry adopts the highest standards”; and (d) the slogan mostly caused Confidence was “Louisiana’s food industry will make sure the seafood you eat is not contaminated”.

Two genders:
How do men and women respond?

Three age groups:
Younger < 30, Middle 31-50, and Older > 51

Four ethnic groups:
White, Black, Hispanic, and Asian

- We learn a lot more by looking at people by the nature of who they are
- And we get a sense of different emotions
- Can the gender (age, ethnic group) differences help us communicate about Louisiana coast seafood safety in a more targeted, effective way?

With Permissions of Howard Moskowitz And iNovum

At the start of a Mind Genomics study, it is necessary to determine a demographic structure of the pollees concerning their gender, age and ethnic group. Depending on the question, type, and the vignette style, different gender (age, ethnic group) may react differently. These differences may be of crucial importance when, in the second phase of Mind Genomics, we approach prospects (prospective customers). Essence of this point of view are: (a) we learn a lot more by looking at people by the nature of who they are; (b) we get a sense of different emotions by looking at the gender (age, ethnic group) they belong to; and (c) gender (age, ethnic group) differences help us communicate a more targeted and a more effective message; in this case, about the Louisiana coast seafood safety.

What makes you feel like you felt last year: by gender

Both genders strong ly affected
Best strategy... admit it, but say generalities, don't focus on the spill...be upbeat

	Males (Constant =28)	
C5	Public needs to realize that imported fish is not well-regulated and MORE likely to be contaminated than Gulf fish	5
B5	*In spite of oil spill. nothing compares with Gulf shrimp. it's healthier than imported shrimp*	-5
	Females (constant=21)	
A4	Over 12,000 oil-related incidents have been reported since 2005 and have been dealt with	7
E4	Louisiana's food industry will make sure the seafood you eat is not contaminated	6
B4	VERY unlikely that ANY oil-contaminated fish will reach your local fishmonger	5
E6	Louisiana's seafood industry also gets its fish from areas not affected by the oil spill	5
B2	*Some people don't care where fish comes from . regardless of the spill*	-5

With Permissions of Howard Moskowitz And iNovum

In the figure corresponding to this paragraph, there is a table with slogans that, within this Mind Genomics study of Louisiana oil spill, mostly turned men/women on and off. For men, (a) the slogans that were the most positive were: "Public needs to realize that imported fish is not well-regulated and MORE likely to be contaminated than Gulf fish", and (b) the most indifferent/negative were: "In spite of oil spill, nothing compares with Gulf shrimp. It's healthier than imported shrimp". For women: (a) the most positive slogans were: "Over 12, 000 oil-related incidents have been reported since 2005 and have been dealt with", "Louisiana's food industry will make sure the seafood you eat is not contaminated", "VERY unlikely that ANY oil-contaminated fish will reach your local fishmonger", and "Louisiana's seafood industry also gets its fish from areas not affected by the oil spill"; (b) the most indifferent/negative slogan for women was "Some people don't care where fish comes from, regardless of the spill".

What elements drive each dominant feeling: by gender

Males...talk social responsibility

Female ...talk scientific proof, and authorities in charge

Males (n=119)	Anx	Cnd	Rel	Cnf
Mother Nature will clear up the effects of the spill	6	5	1	3
Buying seafood from Louisiana will help support families who work in the seafood industry	6	4	5	4
Louisiana's food industry is on top of seafood production standards	5	10	5	6
The federal Centers for Disease Control and Prevention said that tarballs now washing up on beaches pose little danger	4	10	1	3
Oxygen levels are not so low as to be fatal to sea life	3	10	5	1
Oil has a finite lifespan . and then dissipates	3	4	11	1
VERY unlikely that ANY oil-contaminated fish will reach your local fishmonger	1	3	10	6
Louisiana's food industry will make sure the seafood you eat is not contaminated	2	3	7	7
Louisiana's seafood industry employs more than 27,000 people . why should they be punished	1	5	7	6
Louisiana's food industry adopts the highest standards	4	4	7	6
According to environmentalists, less-expensive foreign shellfish -- the other option for fish -- could be far more hazardous to your health than fish from the Gulf	4	4	5	6
Buying seafood from Louisiana will help revive the seafood industry	0	3	4	6
Louisiana's food industry ensures seafood is healthy to eat	1	7	3	6
Females (n=121)	**Anx**	**Cnd**	**Rel**	**Cnf**
The oil spill is not as big of an ecological issue as it's being made out to be	7	0	2	1
You can trust the government to protect the food supply from contamination	7	4	0	4
Louisiana's food industry is on top of seafood production standards	3	9	9	2
Louisiana's food industry ensures seafood is healthy to eat	5	8	5	4
Louisiana's local government will not allow contaminated seafood tp hit the market	5	8	3	6
Oil has a finite lifespan . and then dissipates	4	8	4	2
Louisiana's seafood industry also gets its fish from areas not affected by the oil spill	4	8	7	4
Louisiana's food industry adopts the highest standards	2	9	10	0
Public needs to realize that imported fish is not well-regulated and MORE likely to be contaminated than Gulf fish	5	6	8	1
Shrimp and crab are abundant and safe to eat	3	4	8	1
The federal Centers for Disease Control and Prevention said that tarballs now washing up on beaches pose little danger	3	0	8	3
Laboratory instruments can detect chemical contamination of seafood & prevent it from being sold	1	7	7	4
Over 12,000 oil-related incidents have been reported since 2005 and have been dealt with	-1	2	5	6

With Permissions of Howard Moskowitz And iNovum

Talking about emotions induced by the slogans, men talk about responsibility and women about scientific proof and authorities in charge. Among men, the slogans mostly producing emotions were: (a) Anxiousness "Mother Nature will clear up the effects of the spill", (b) Concern "Louisiana's food industry is on top of seafood production standards", (c) Relief "Oil has a finite lifespan … and then dissipates", and (d) Confidence "Louisiana's food industry will make sure the seafood you eat is not contaminated". For women, the slogans mostly producing emotions were: (a) Anxiousness "The oil spill is not as big of an ecological issue as it's being made out to be", (b) Concern "Louisiana's food industry is on top of seafood production standards", (c) Relief "Louisiana's food industry adopts the highest standards", and (d) Confidence "Over 12,000 oil-related incidents have been reported since 2005 and have been dealt with".

What makes you feel like you felt last year: by age group

Younger…talk facts
Older…talk standards and authorities in charge

		Y	M	O
	Youngest Age<30 (n=51, constant = 30)			
C5	Public needs to realize that imported fish is not well-regulated and MORE likely to be contaminated than Gulf fish	8	3	5
A4	Over 12,000 oil-related incidents have been reported since 2005 and have been dealt with	6	3	5
A1	*The oil spill is not as big of an ecological issue as it's being made out to be*	-6	-2	-2
C3	*Government agencies err on the side of caution when it comes to food safety*	-6	2	1
F6	*Buy seafood from Louisiana . showing friends and family that you care*	-7	2	2
F5	*It is your duty to support Louisiana's seafood industry over imported seafood*	-8	2	-1
A2	*Louisiana's seafood industry employs more than 27,000 people . why should they be punished*	-11	-2	3
	Middle: Age<31-50 (n=93, constant = 17)			
E4	Louisiana's food industry will make sure the seafood you eat is not contaminated	2	5	4
B2	*Some people don't care where fish comes from . regardless of the spill*	-4	-5	-5
	Oldest: Age>50 (n=96, constant = 28)			
E6	Louisiana's seafood industry also gets its fish from areas not affected by the oil spill	-1	2	6
E3	Louisiana's food industry is on top of seafood production standards	2	3	6
E1	Louisiana's food industry adopts the highest standards	-3	2	5
A4	Over 12,000 oil-related incidents have been reported since 2005 and have been dealt with	6	3	5
C5	Public needs to realize that imported fish is not well-regulated and MORE likely to be contaminated than Gulf fish	8	3	5
B4	VERY unlikely that ANY oil-contaminated fish will reach your local fishmonger	0	2	5
B2	*Some people don't care where fish comes from . regardless of the spill*	-4	-5	-5
B5	*In spite of oil spill. nothing compares with Gulf shrimp. it's healthier than imported shrimp*	4	1	-6

21

With Permissions of Howard Moskowitz And iNovum

In this figure, there is a table describing how different age groups reacted to slogans used on vignettes in the Louisiana oil spill study. For the youngest age group (<30 years), (a) the most positive were "Public needs to realize that imported fish is not well-regulated and MORE likely to be contaminated than Gulf fish" and "Over 12,000 oil-related incidents have been reported since 2005 and have been dealt with, and (b) the most indifferent/negative were "In spite of oil spill, nothing compares with Gulf shrimp. It's healthier than imported shrimp" and "Louisiana's seafood industry employs more than 27,000 people … why should they be punished". For the middle age group (>30 and <50 years), (c) the most positive slogans were "Louisiana's food industry will make sure the seafood you eat is not contaminated", and (d) the most indifferent/negative slogan was "Some people don't care where fish comes from … regardless of the spill". For the older age group (>50 years), (e) the most positive slogans was "Louisiana's seafood industry also gets its fish from areas not affected by the oil spill", and (f) the most indifferent/negative slogan was "In spite of oil spill … nothing compares with Gulf shrimp. It's healthier than imported shrimp".

What elements drive each dominant feeling: by age group

Younger respondents show the strongest emotions.
Older respondents are concerned.

Under 30	Anx	Cnd	Rel	Cnf
The oil spill is not as big of an ecological issue as it's being made out to be	13	0	2	2
Oxygen levels are not so low as to be fatal to sea life	-3	12	3	2
Louisiana's food industry is on top of seafood production standards	2	11	5	1
People in the fishing industry are afraid of long-term damage to their livelihood if people stop buying	9	12	-2	3
Safety ensured. oil-contaminated shrimp would smell like oil. danger would be obvious	2	10	-5	6
It will take time for exposure to oil to cease. once it does, fish eliminate contaminants quickly	-2	10	2	3
The Gulf has natural "oil eating" bacteria	6	1	14	5
Over 12,000 oil-related incidents have been reported since 2005 and have been dealt with	3	-5	12	4
Louisiana's food industry adopts the highest standards	-1	6	11	0
The federal Centers for Disease Control and Prevention said that exposure to small amounts of oil and dispersants is not harmful	1	-4	11	3
It is your duty to support Louisiana's seafood industry over imported seafood	0	8	11	0
Americans' appetite for seafood won't be changed overnight	-1	1	11	2
Age 31-50	**Anx**	**Cnd**	**Rel**	**Cnf**
Oil has a finite lifespan . and then dissipates	1	11	8	1
Louisiana's food industry will make sure the seafood you eat is not contaminated	5	3	12	5
Louisiana's food industry adopts the highest standards	3	5	13	3
Louisiana's food industry ensures seafood is healthy to eat	0	9	-1	10
Age 50+	**Anx**	**Cnd**	**Rel**	**Cnf**
Louisiana's food industry is on top of seafood production standards	4	10	6	4
Louisiana's local government will not allow contaminated seafood tp hit the market	4	10	5	4

With Permissions of Howard Moskowitz And iNovum

Talking about emotions induced by the slogans, younger respondents showed stronger different emotions while older were mostly showing concern. Among the younger age group, the slogan mostly causing (a) Anxiousness was "The oil spill is not as big of an ecological issue as it's being made out to be", (b) Concern "Oxygen levels are not so low as to be fatal to sea life", (c) Relief "Oxygen levels are not so low as to be fatal to sea life", For the middle age group, the slogan mostly causing (a) Concern was "Oil has a finite lifespan, and then dissipates", (b) Relief "Louisiana's food industry will make sure the seafood you eat is not contaminated", and (c) Confidence "Louisiana's food industry ensures seafood is healthy to eat". For the older age group, the slogans mostly causing Concern were "Louisiana's food industry is on top of seafood production standards" and "Louisiana's local government will not allow contaminated seafood to hit the market".

What makes you feel like you felt last year: by ethnic group

Stress standards.
Different ingoing biases predilections (Hispanic & Asian most doubtful)

	White (n=62, constant = 33)	
E4	Louisiana's food industry will make sure the seafood you eat is not contaminated	8
D6	The Gulf has natural "oil eating" bacteria	6
D5	*Mother Nature will clear up the effects of the spill*	-6
B2	*Some people don't care where fish comes from . regardless of the spill*	-7
A1	*The oil spill is not as big of an ecological issue as it's being made out to be*	-6
	Black (n=67, constant = 27)	
E5	Louisiana's local government will not allow contaminated seafood tp hit the market	10
B4	VERY unlikely that ANY oil-contaminated fish will reach your local fishmonger	7
E6	Louisiana's seafood industry also gets its fish from areas not affected by the oil spill	6
E2	Louisiana's food industry ensures seafood is healthy to eat	6
A2	*Louisiana's seafood industry employs more than 27,000 people . why should they be punished*	-6
	Hispanic (n=42, constant = 18)	
C5	Public needs to realize that imported fish is not well-regulated and MORE likely to be contaminated than Gulf fish	10
A5	The federal Centers for Disease Control and Prevention said that tarballs now washing up on beaches pose little danger	9
A3	Spill was out in the open ocean. less damaging than if it were closer to the coast	9
D3	Oxygen levels are not so low as to be fatal to sea life	9
F5	It is your duty to support Louisiana's seafood industry over imported seafood	8
A4	Over 12,000 oil-related incidents have been reported since 2005 and have been dealt with	8
D6	The Gulf has natural "oil eating" bacteria	7
F6	Buy seafood from Louisiana . showing friends and family that you care	7
A6	The federal Centers for Disease Control and Prevention said that exposure to small amounts of oil and dispersants is not harmful	6
	Asian (n=69, constant = 19)	
C5	Public needs to realize that imported fish is not well-regulated and MORE likely to be contaminated than Gulf fish	10
E1	Louisiana's food industry adopts the highest standards	8
B2	*Some people don't care where fish comes from . regardless of the spill*	-6
F5	*It is your duty to support Louisiana's seafood industry over imported seafood*	-7

With Permissions of Howard Moskowitz And iNovum

In the figure corresponding to this paragraph, there is a table describing how different ethnic groups reacted to the slogans. For the white ethnic group: (a) the most positive was "Louisiana's food industry will make sure the seafood you eat is not contaminated", and (b) the most indifferent/negative was "The oil spill is not as big ecological issue as it is being made out to be". For the black ethnic group: (c) the most positive slogans was "Louisiana's local government will not allow contaminated seafood to hit the market", and (d) the most indifferent/negative slogan was "Louisiana's seafood industry employs more than 27,000 people … why should they be punished". For the Hispanic ethnic group: (e) the most positive slogans was "Public needs to realize that imported fish is not well-regulated and MORE likely to be contaminated than Gulf fish", and (f) the most indifferent/negative slogan was "The federal Centers for Disease Control and Prevention said that exposure to small amounts of oil and dispersants is not harmful". Finally, for the Asian ethnic group: (g) the most positive slogans was "Public needs to realize that imported fish is not well-regulated and MORE likely to be contaminated than Gulf fish", and (h) the most indifferent/negative slogan was "It is your duty to support Louisiana's seafood industry over imported seafood".

What elements drive each dominant feeling: by ethnic group

White (n=62)	Anx	Cnd	Rel	Cnf
The oil spill is not as big of an ecological issue as it's being made out to be	9	4	-1	3
The loyalty of local consumers will get Louisiana fishermen through the fallout from the spill	8	3	0	8
Louisiana's food industry is on top of seafood production standards	-1	15	6	4
Louisiana's local government will not allow contaminated seafood tp hit the market	3	13	0	3
Laboratory instruments can detect chemical contamination of seafood & prevent it from being sold	3	11	4	1
Oil has a finite lifespan . and then dissipates	4	10	5	1
Public needs to realize that imported fish is not well-regulated and MORE likely to be contaminated than Gulf fish	1	10	6	2
According to environmentalists, less-expensive foreign shellfish -- the other option for fish -- could be far more hazardous to your health than fish from the Gulf	-1	10	4	4
Oxygen levels are not so low as to be fatal to sea life	-3	10	7	1
Louisiana's food industry adopts the highest standards	1	10	10	4
Shrimp and crab are abundant and safe to eat	0	7	9	2
Louisiana's seafood industry also gets its fish from areas not affected by the oil spill	1	0	7	11
Louisiana's food industry will make sure the seafood you eat is not contaminated	0	5	5	10
Buying seafood from Louisiana will help support families who work in the seafood industry	7	3	3	10

Black (n=67)	Anx	Cnd	Rel	Cnf
People in the fishing industry are afraid of long-term damage to their livelihood if people stop buying seafood	7	13	1	0
Louisiana's seafood industry also gets its fish from areas not affected by the oil spill	1	14	2	4
In spite of oil spill. nothing compares with Gulf shrimp. it's tastierthan imported shrimp	2	12	1	1
Louisiana's food industry ensures seafood is healthy to eat	1	12	0	6
The Gulf has natural "oil eating" bacteria	2	11	4	2
The federal Centers for Disease Control and Prevention said that exposure to small amounts of oil and dispersants is not harmful	2	7	11	0
The federal Centers for Disease Control and Prevention said that tarballs now washing up on beaches pose little	1	6	10	1
Buying seafood from Louisiana will help support families who work in the seafood industry	4	0	10	2
Louisiana's local government will not allow contaminated seafood to hit the market	5	8	5	8

With Permissions of Howard Moskowitz And iNovum

In the figure corresponding to this paragraph, there is a table describing what emotions the slogans produced in representatives of white and black ethnic groups from the polled population. Among the white ethnic group, the slogan mostly causing (a) Anxiousness was "The oil spill is not as big of an ecological issue as it's being made out to be", (b) Concern "Louisiana's food industry is on top of seafood production standards", (c) Relief "Louisiana's food industry adopts the highest standards" and (d) Confidence "Louisiana's seafood industry also gets its fish from areas not affected by the oil spill". For the black ethnic group, the slogan mostly causing (a) Anxiousness was “People in the fishing industry are afraid of long-term damage to their livelihood if people stop buying seafood”, (b) Concern “Louisiana’s seafood industry also gets its fish from areas not affected by the oil spill”, (c) Relief “The federal Centers for Disease Control and Prevention said that exposure to small amounts of oil and dispersants is not harmful”, and (d) Confidence “Louisiana’s local government will not allow contaminated seafood to hit the market”.

What elements drive each dominant feeling: by ethnic group

	Anx	Cnd	elieve	Cnf
Hispanic (n=43)			0	
Buy seafood from Louisiana . showing friends and family that you care	13	0	2	3
The oil spill is not as big of an ecological issue as it's being made out to be	11	-1	0	2
Buying seafood from Louisiana will help support families who work in the seafood industry	4	11	4	1
Louisiana's seafood industry also gets its fish from areas not affected by the oil spill	2	-3	14	3
VERY unlikely that ANY oil-contaminated fish will reach your local fishmonger	2	5	11	5
Oil has a finite lifespan . and then dissipates	4	4	10	2
It will take time for exposure to oil to cease. once it does, fish eliminate contaminants quickly	2	4	10	1
Louisiana's food industry will make sure the seafood you eat is not contaminated	4	3	10	2
In spite of oil spill. nothing compares with Gulf shrimp. it's tastier than imported shrimp	6	1	10	4
Mother Nature will clear up the effects of the spill	3	3	2	7
Asian (n=68)				
Shrimp and crab are abundant and safe to eat	10	-2	5	7
Oxygen levels are not so low as to be fatal to sea life	10	6	4	1
Louisiana's seafood industry also gets its fish from areas not affected by the oil spill	8	10	6	0
Louisiana's seafood industry employs more than 27,000 people . why should they be punished	5	10	0	4
In spite of oil spill. nothing compares with Gulf shrimp. it's healthier than imported shrimp	-2	6	13	2
Laboratory instruments can detect chemical contamination of seafood & prevent it from being sold	-2	4	13	7
Government agencies err on the side of caution when it comes to food safety	5	-2	11	3
Americans' appetite for seafood won't be changed overnight	2	-1	10	1
Public needs to realize that imported fish is not well-regulated and MORE likely to be contaminated than Gulf fish	7	-1	8	7
According to environmentalists, less-expensive foreign shellfish -- the other option for fish -- could be far more hazardous to your health than fish from the Gulf	8	3	6	7

With Permissions of Howard Moskowitz And iNovum

In the figure corresponding to this paragraph, there is a table describing what emotions the slogans produced in representatives of Hispanic and Asian ethnic groups from the polled population. Among the Hispanic ethnic group, the slogan mostly causing (a) Anxiousness was “Buy seafood from Louisiana … showing friends and family that you care”, (b) Concern “Buying seafood from Louisiana will help support families who work in the seafood industry”, (c) Relief “Louisiana’s seafood industry also gets its fish from areas not affected by the oil spill” and (d) Confidence “Mother Nature will clear up the effects of the spill”. For the Asian ethnic group, the slogan mostly causing (a) Anxiousness was “Shrimp and crab are abundant and safe to eat”, (b) Concern “Louisiana’s seafood industry also gets its fish from areas not affected by the oil spill”, (c) Relief “In spite of oil spill … nothing compares with Gulf shrimp. It’s healthier than imported shrimp”, and (d) Confidence “Public needs to realize that imported fish is not well-regulated and MORE likely to be contaminated than Gulf fish”.

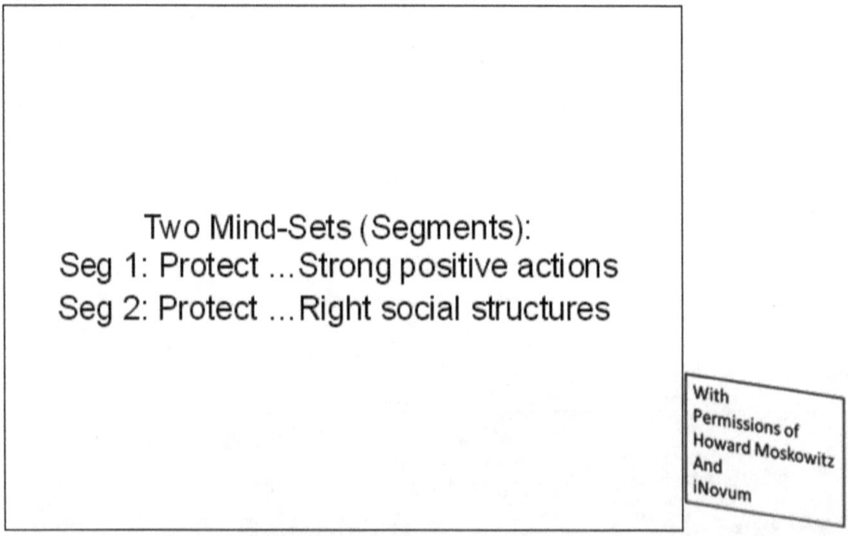

Mind Genomics does not offer recipes, it offers guidelines. This means the experience of Mind Genomics experts plays an important role. Consequently, the success of a particular case study process depends strongly on the experience of the expert. In the Louisiana oil spill case, an experienced expert would reveal two mind types. The first one would characterize people as those who emphasize strong positive actions, think that the nature gives alternatives, like to take charge, and believe they can do it. The second mind type would characterize people as those who emphasize that the society has right structures in place to protect its members, who want to know that all is OK, that the nature isn't so bad, and who get very nervous when they hear science talk.

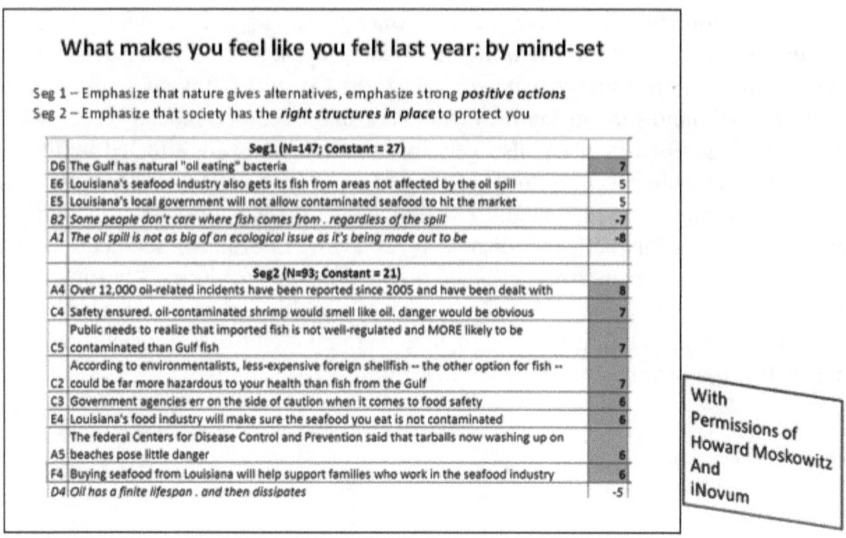

What makes you feel like you felt last year: by mind-set

Seg 1 – Emphasize that nature gives alternatives, emphasize strong ***positive actions***
Seg 2 – Emphasize that society has the ***right structures in place*** to protect you

	Seg1 (N=147; Constant = 27)	
D6	The Gulf has natural "oil eating" bacteria	7
E6	Louisiana's seafood industry also gets its fish from areas not affected by the oil spill	5
E5	Louisiana's local government will not allow contaminated seafood to hit the market	5
B2	*Some people don't care where fish comes from . regardless of the spill*	-7
A1	*The oil spill is not as big of an ecological issue as it's being made out to be*	-8
	Seg2 (N=93; Constant = 21)	
A4	Over 12,000 oil-related incidents have been reported since 2005 and have been dealt with	8
C4	Safety ensured. oil-contaminated shrimp would smell like oil. danger would be obvious	7
C5	Public needs to realize that imported fish is not well-regulated and MORE likely to be contaminated than Gulf fish	7
C2	According to environmentalists, less-expensive foreign shellfish -- the other option for fish -- could be far more hazardous to your health than fish from the Gulf	7
C3	Government agencies err on the side of caution when it comes to food safety	6
E4	Louisiana's food industry will make sure the seafood you eat is not contaminated	6
A5	The federal Centers for Disease Control and Prevention said that tarballs now washing up on beaches pose little danger	6
F4	Buying seafood from Louisiana will help support families who work in the seafood industry	6
D4	*Oil has a finite lifespan . and then dissipates*	-5

In the figure corresponding to this paragraph, there is a table presenting the slogans of choice for the both mind types. The numbers in the right-most column tell how much a slogan was liked, found indifferent, or disliked. For the mind type one (Segment #1), the preferred slogans were: (a) "Gulf has natural 'oil eating' bacteria", "Louisiana's seafood industry also gets its fish from areas not affected by the oil spill", and "Louisiana's local government will not allow contaminated seafood to hit the market". The mostly disliked slogan was "The oil spill is not as big ecological issue as it is being made out to be". For the Segment #2, the preferred slogans were: (a) "Over 12,000 oil-related incidents have been reported since 2005 and have been dealt with", "Safety ensured. Oil-contaminated shrimp would smell like oil. Danger would be obvious", and "Public needs to realize that imported fish is not well-regulated and MORE likely to be contaminated than Gulf fish". The mostly disliked slogan was "Oil has a finite lifespan … and then dissipates".

What elements drive each dominant feeling : Seg1 - Action

They want to take charge, and believe they can

Seg1 (n=147)	Anx	Cnd	Rel	Cnf
The oil spill is not as big of an ecological issue as it's being made out to be	5	-1	0	2
Oil has a finite lifespan . and then dissipates	5	8	9	1
Mother Nature will clear up the effects of the spill	5	-2	8	4
Louisiana's food industry is on top of seafood production standards	5	12	8	4
people stop buying seafood	3	9	4	1
Louisiana's food industry ensures seafood is healthy to eat	2	8	4	5
Oil has a finite lifespan . and then dissipates	5	8	9	1
Louisiana's food industry adopts the highest standards	3	6	9	3
VERY unlikely that ANY oil-contaminated fish will reach your local fishmonger	3	2	9	2
Louisiana's seafood industry also gets its fish from areas not affected by the oil spill	2	7	8	5
Laboratory instruments can detect chemical contamination of seafood & prevent it from	0	7	8	3
Oxygen levels are not so low as to be fatal to sea life	3	6	8	1
It will take time for exposure to oil to cease. once it does, fish eliminate contaminants	2	2	8	4
Mother Nature will clear up the effects of the spill	5	-2	8	4
Louisiana's seafood industry also gets its fish from areas not affected by the oil spill	2	7	8	5
Buying seafood from Louisiana will help support families who work in the seafood	3	4	7	5
Louisiana's food industry will make sure the seafood you eat is not contaminated	3	5	6	5
Louisiana's food industry ensures seafood is healthy to eat	2	8	4	5
Louisiana's local government will not allow contaminated seafood tp hit the market	4	7	3	5

With Permissions of Howard Moskowitz And iNovum

In the figure corresponding to this paragraph, there is a table describing what emotions the slogans produced in pollees grouped into Segment #1. For them, slogans mostly causing (a) Anxiousness were "The oil spill is not as big of an ecological issue as it's being made out to be" and "Oil has a finite lifespan … and then dissipates", (b) Concern "Louisiana's food industry is on top of seafood production standards" and "Louisiana's food industry ensures seafood is healthy to eat", (c) Relief "Louisiana's food industry adopts the highest standards "and "VERY unlikely that ANY oil-contaminated fish will reach your local fishmonger", and (d) Confidence "Louisiana's local government will not allow contaminated seafood to hit the market" and "Louisiana's food industry will make sure the seafood you eat is not contaminated".

What elements drive each dominant feeling: Seg2 - Structure

They want to know it's all ok
They want to know that nature isn't so bad
They get very nervous when they hear science talk

Seg2 (n=93)	Anx	Cnd	Rel	Cnf
The federal Centers for Disease Control and Prevention said that exposure to small amounts of oil and dispersants is not harmful	9	9	3	2
Safety ensured. oil-contaminated shrimp would smell like oil. danger would be obvious	8	2	2	2
Louisiana's local government will not allow contaminated seafood tp hit the market	5	10	7	4
The oil spill is not as big of an ecological issue as it's being made out to be	9	9	6	1
The federal Centers for Disease Control and Prevention said that exposure to small amounts of oil and dispersants is not harmful	9	9	3	2
It will take time for exposure to oil to cease. once it does, fish eliminate contaminants	4	9	2	-1
You can trust the government to protect the food supply from contamination	5	8	2	5
Oxygen levels are not so low as to be fatal to sea life	2	8	0	4
Laboratory instruments can detect chemical contamination of seafood & prevent it from	-1	8	5	5
Spill was out in the open ocean. less damaging than if it were closer to the coast	-2	8	6	2
Shrimp and crab are abundant and safe to eat	1	6	8	1
It is your duty to support Louisiana's seafood industry over imported seafood	1	4	8	1
VERY unlikely that ANY oil-contaminated fish will reach your local fishmonger	5	1	5	8

With Permissions of Howard Moskowitz And iNovum

In the figure corresponding to this paragraph, there is a table describing what emotions the slogans produced in pollees grouped into Segment #2. For them, slogans mostly causing Anxiousness were "The federal Centers for Disease Control and Prevention said that exposure to small amounts of oil and dispersants is not harmful" and "Safety ensured. Oil-contaminated shrimp would smell like oil. danger would be obvious", (b) Concern "Louisiana's local government will not allow contaminated seafood to hit the market" and "The oil spill is not as big of an ecological issue as it's being made out to be", (c) Relief "Shrimp and crab are abundant and safe to eat" and "It is your duty to support Louisiana's seafood industry over imported seafood", and (d) Confidence "VERY unlikely that ANY oil-contaminated fish will reach your local fishmonger".

The segments' demographic patterns: You can't discover one

We have clearly different segments ...at least in the mind
They should be traceable to the subgroups
No...they are not... segments exist...but they transcend standard classification

	Action Seg.1	Structure Seg.2
Total	61%	39%
Males	59	41
Females	64	36
Age < 30	55	45
Age 31-50	63	37
Age > 50	62	38
White	63	37
Black	61	39
Hispanic	58	42
Asian	62	38

With Permissions of Howard Moskowitz And iNovum

In the figure corresponding to this paragraph, there is a table describing how demographic structures correspond to discovered mind types. The data in the table is proving what has been said before "the demography is not a good base to rely upon in creating a marketing campaign". One can see that all demographic groups are almost evenly split between the segments, though the segment #1 is larger in all of them: (a) In males, the ratio of segment #1 to segment #2 is 59 to 41, in females 64 to 36; (b) with the young age group it is 55 to 45, with middle age 63 to 37, and with older age 62 to 38; and (c) in the white group, the ratio is 63 to 37, in the black ethnic group it is 61 to 39, in the Hispanic ethnic group it is 58 to 42, and in the Asian ethnic group it is 62 to 38.

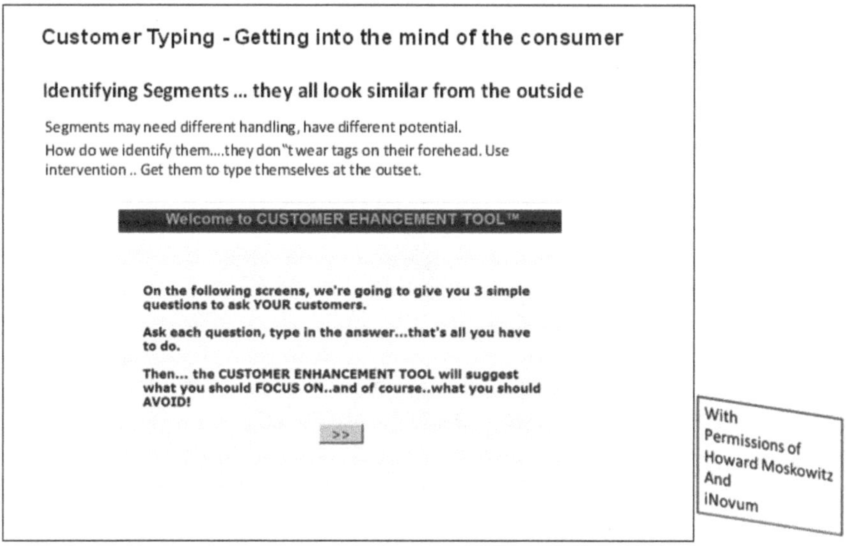

This paragraph is related to this and the next three figures. Customer Typing is a short process of determining to which mind type the customer belongs. As said before, in the Internet scenario, the customer typing is done after the customer finished the initially planned activity. In this Mind Genomics study the customer was asked to express how comfortable (s)he was about buying/eating Louisiana shrimp after reading three different slogans (as presented in the next three figures). Possible feelings could be rated with "1—Very uncomfortable", "2—Not sure", and "About as comfortable as a year ago". The three slogans were "Government agencies err on the side of caution when it comes to food safety", "Oil has a finite life span … and then dissipates", and "In spite of oil spill … nothing compares to Gulf shrimp … it's tastier than imported shrimp". If the customer circled "1" two or three times, s(he) belonged to the first mind type segment.

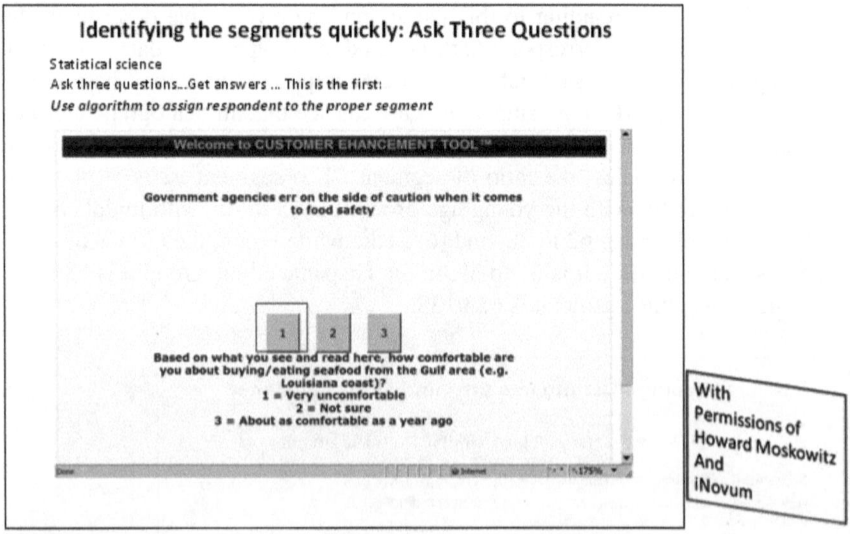
Identifying the segments quickly: Ask Three Questions
Statistical science
Ask three questions...Get answers ... This is the first:
Use algorithm to assign respondent to the proper segment
Welcome to CUSTOMER EHANCEMENT TOOL™
Government agencies err on the side of caution when it comes to food safety
1
2
3
Based on what you see and read here, how comfortable are you about buying/eating seafood from the Gulf area (e.g. Louisiana coast)?
1 = Very uncomfortable
2 = Not sure
3 = About as comfortable as a year ago
With Permissions of Howard Moskowitz And iNovum

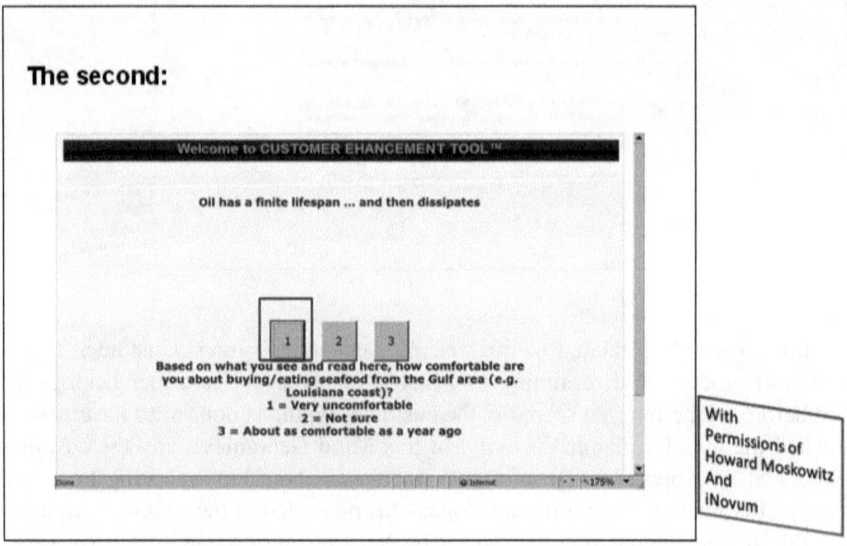
The second:
Welcome to CUSTOMER EHANCEMENT TOOL™
Oil has a finite lifespan ... and then dissipates
1
2
3
Based on what you see and read here, how comfortable are you about buying/eating seafood from the Gulf area (e.g. Louisiana coast)?
1 = Very uncomfortable
2 = Not sure
3 = About as comfortable as a year ago
With Permissions of Howard Moskowitz And iNovum

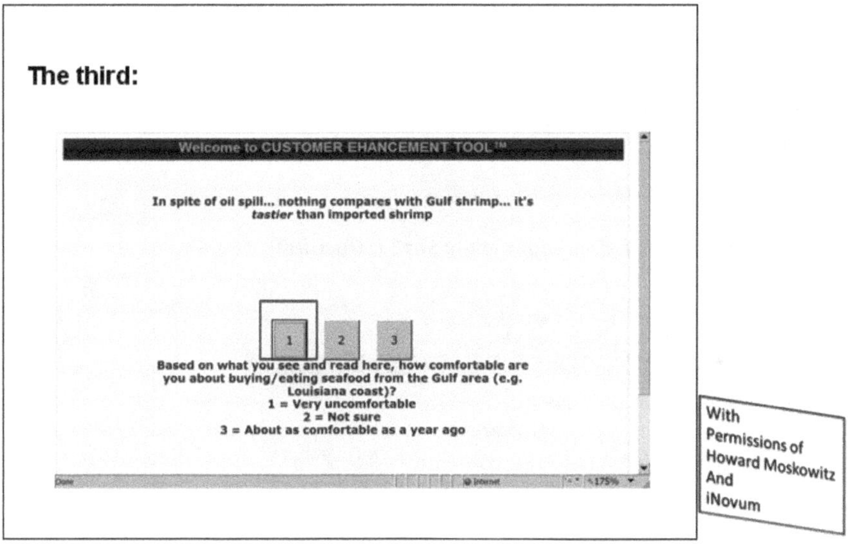

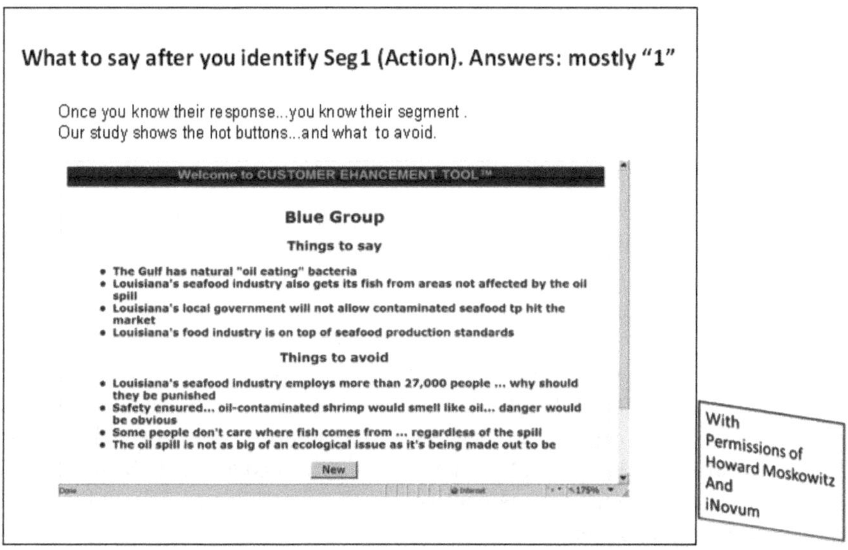

After the customer is typed to belong to the first segment (nicknamed "Blue Group"), these are the thing to tell him/her: (a) "The Gulf has natural "oil eating" bacteria", "Louisiana's seafood industry also gets its fish from areas not affected by the oil spill", "Louisiana's local government will not allow contaminated seafood to hit the market", and "Louisiana's food industry is on top of seafood production standards". The following statements should be avoided when talking to the first mind type: (b) "Louisiana's seafood industry employs more than 27,000 people … why should they be punished", "Safety ensured… oil-contaminated shrimp would smell like oil… danger would be obvious", "Some people don't care where fish comes from … regardless of the spill", and "The oil spill is not as big of an ecological issue as it's being made out to be".

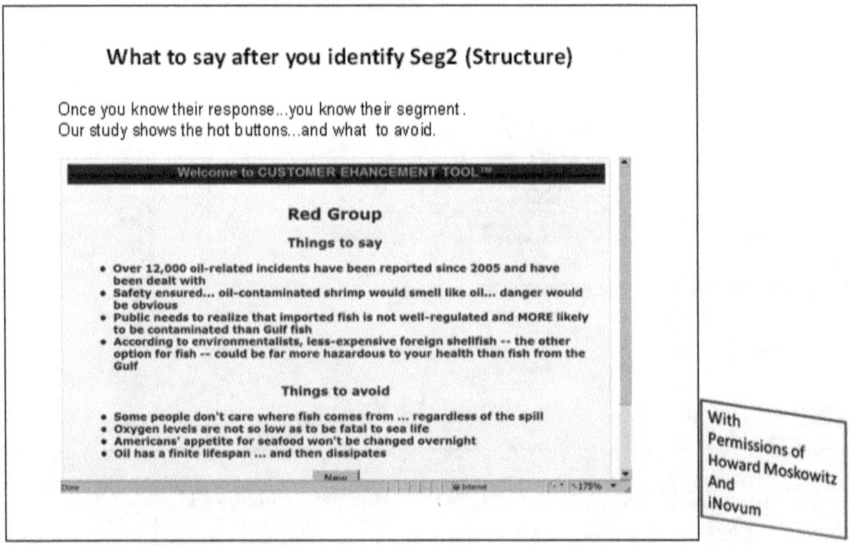

After the customer is typed to belong to the second segment (nicknamed "Red Group"), these are the thing to tell him/her: (a) "Over 12,000 oil-related incidents have been reported since 2005 and have been dealt with", "Safety ensured… oil-contaminated shrimp would smell like oil… danger would be obvious", "Public needs to realize that imported fish is not well-regulated and MORE likely to be contaminated than Gulf fish", and "According to environmentalists, less-expensive foreign shellfish—the other option for fish—could be far more hazardous to your health than fish from the Gulf". The following statements should be avoided when talking to the second mind type: (b) "Some people don't care where fish comes from … regardless of the spill", "Oxygen levels are not so low as to be fatal to sea life", "Americans' appetite for seafood won't be changed overnight", and "Oil has a finite lifespan … and then dissipates".

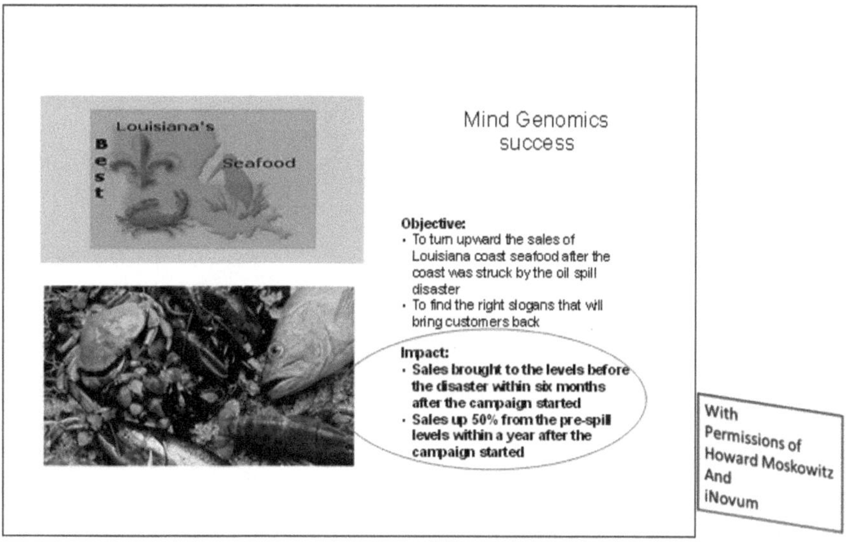

In conclusion, the figure corresponding to this paragraph summarizes the major objective placed before Mind Genomics in this particular case, as well as the major impact that it was able to generate. The major two objectives were: (a) To turn upward the sales of Louisiana coast seafood after the coast was struck by the oil spill disaster and (b) To find the right slogans that will bring customers back. The major two impacts were: (a) Sales brought to the levels before the disaster within six months after the campaign started and (b) Sales up 50 % from the pre-spill levels within a year after the campaign started. The bottom line is that mathematics combined with psychology can create great marketing results.

3.3 Case #3—Higher Education Student Enrolment

As it has been said, Mind Genomics can be used in many, many fields. This presentation leads through the New York City Queens College applications process. The goal was to improve the conversion of applicants to enrollment ratios, to secure better retention and graduation rates, and to reduce lead general costs. The process determined three ViewPoint mind segments (mind types): "Opportunistic Knowledge Seekers" who are interested in a variety of academic programs and flexible schedules, "Motivated Go Getters" who are Interested in continuing education at top grad schools and award-winning faculties, and "Mentor Me" who are Interested in freshmen programs to adjust to college life and are not concerned with other academic offerings. Carefully chosen motivational slogans, appropriate for each one of the mind-set groups, succeeded in increasing the enrollment by 30 %.

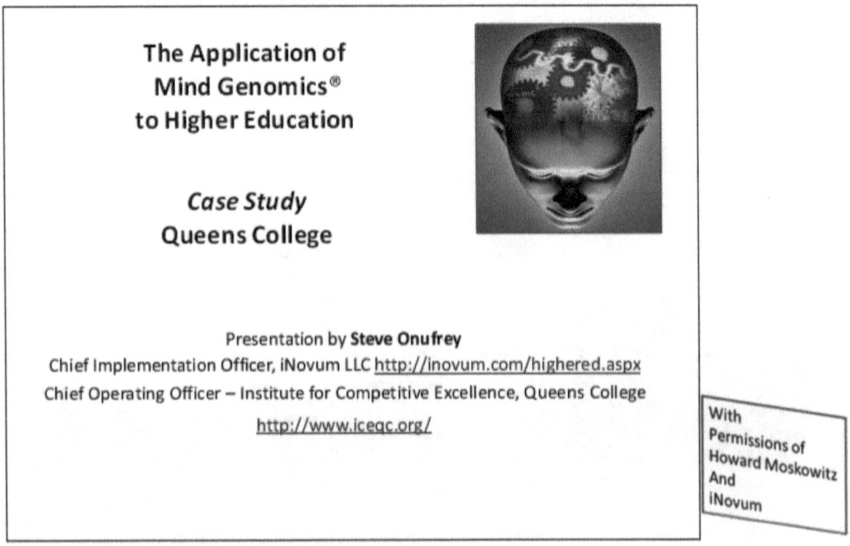

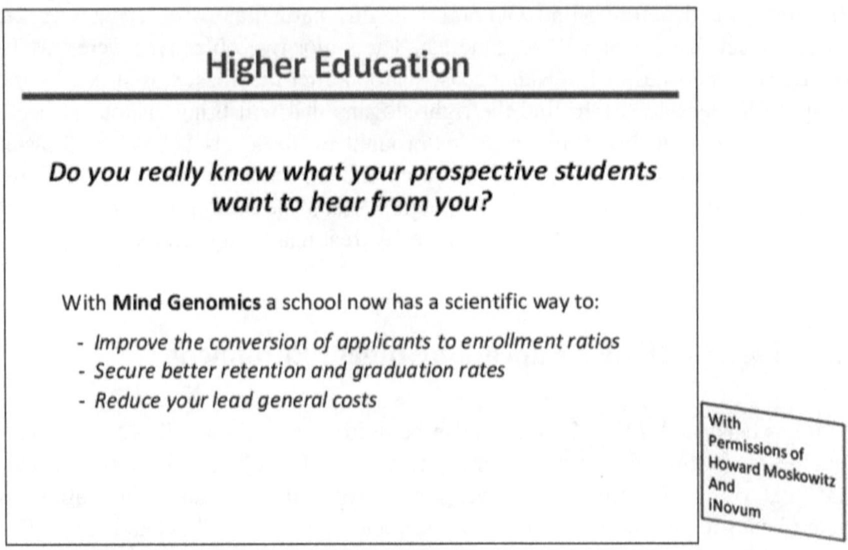

All colleges in the world, from time to time, generate new education programs aimed at helping some important mission of the society and/or giving their students new important opportunities. No matter how great a new program is, there is always some inertia in its acceptance by the society. Mind Genomics proves to be able to help accelerate the acceptance process. The acceptance process is accelerated if educators know what their prospective students want to hear from them.

The same holds for the existing courses that need more enrollments. To make a long story short, the answers that this Mind Genomics study submitted respond to the following three basic questions: (a) how to improve the conversion of applicants to enrollment ratios, (b) how to secure better retention and graduation rates, and (c) how to reduce general lead costs.

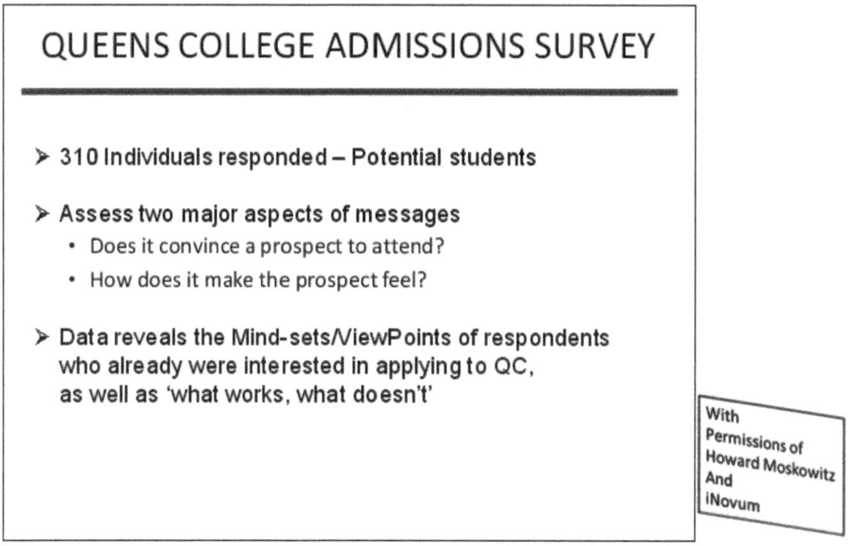

In order to find effective answers to the above mentioned three crucial questions, a poll was organized for potential students. 310 responded to the poll what was enough for an effective Mind Genomics study. Slogans were generated to reveal two mayor aspects: (a) how convenient the program is, and (b) how the prospective student feels about it. For this study, the most relevant is the ViewPoint (mind type/mind set) of the respondents. In another words, those who visited the enrollment web site of the Queens College (QC), with an intention to select a program, had to be classified first, based on the issue which proves to be the most important for a study of this sort: ViewPoint. Of course, it is difficult to know ahead of time what ViewPoints will pop-up as the most frequent ones.

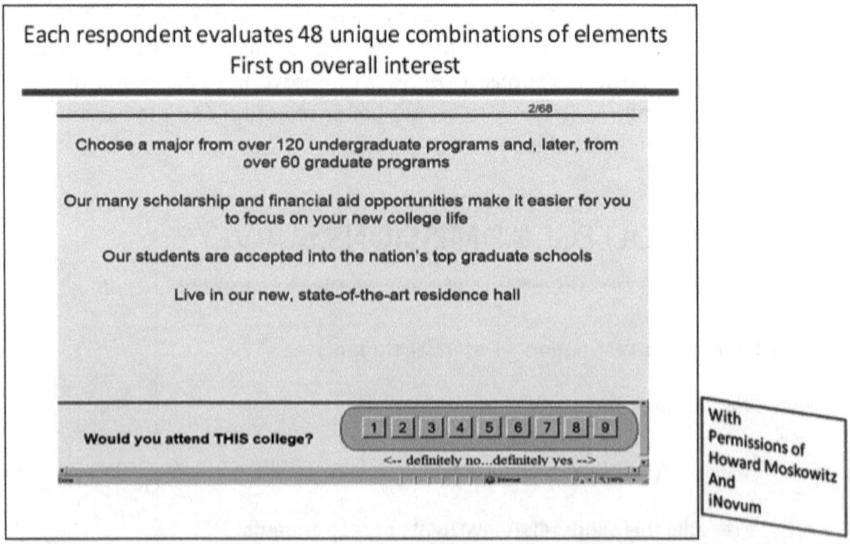

For this study, 48 vignettes were generated, using the method described before. One example of a vignette is visible on two figures related to this paragraph, one next to the paragraph, and one after it. The only difference is in the question related to the vignette. In the first case the question was "Would you attend this college?". The answer was expected on the relative scale, from 1 (definitely no) to 9 (definitely yes). Note that absolute answers could only be contra-productive in Mind Genomics studies. This is because, for example, $100 is a lot for one person and a little for another. What brings value to the study are relative numbers, like larger or smaller, more or less, etc. In the second case, the question is related to the feelings one gets after reading the slogans—one single emotion: 1—Eager, 2—Uncertain, 3—Comfortable, 4—Intimidated, and 5—Curious.

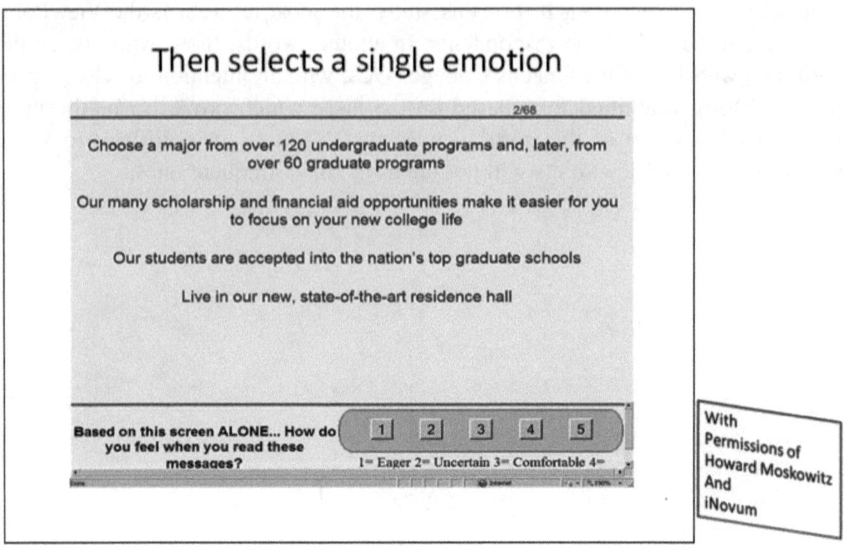

The total panel's interest is different from that in each of the three identified segments

1) Would you attend THIS college? <-- Definitely NO Definitely YES --> 1 2 3 4 5 6 7 8 9 Sorted by Total Sample : Highlighted >+9 winners & <-9 losers	Total Sample	Seg1 of 3 Knowledge	Seg2 of 3 Ambitions	Seg3 of 3 Help me to start
Base Size	310	175	61	74
Constant	51	49	63	47
Our students are accepted into the nation's top graduate schools	9	7	17	6
We help students make realistic connections between education and the world beyond	8	10	-1	11
We're one of only two colleges in the U.S. to offer study abroad opportunities to freshmen	7	3	10	15
Freshman Year Initiative helps you make a smooth transition from high school to college	7	6	5	13
Take advantage of our honors programs in the arts; sciences; or social sciences	7	11	14	-9
Study and work on projects with our award-winning faculty	7	5	12	7
Choose a major from over 120 undergraduate programs and, late[illegible] graduate programs	7	12	5	-4
Our college is minutes away from Manhattan-you can even see the skyline from our Quad	7	1	8	19
Learn job-seeking skills and build your resume	7	6	0	13
Flexible schedules: You can attend classes part time or full time, day or night, and even on weekends	6	12	-4	2
We offer internships in many fields, from the media to finance, as well as job-placement services	6	5	0	14
We offer the best value for your money-and a great education!	6	11	7	-5
We offer a Bachelor of Business Administration with three areas of concentration and a unique minor in business and the liberal arts	1	-4	8	5
We have new sports fields and the latest exercise equipment	0	-4	-2	10

With Permissions of Howard Moskowitz And iNovum

As indicated in the figure corresponding to this paragraph, the survey was performed and, on the whole polled population, it showed that respondents are interested in school prestige and career opportunities and are not concerned with athletic facilities. Also, the total panel's interest is different from the ones in each of the three identified segments. The most appreciated slogans for all pollees together were: "Our students are accepted into the nation's top graduate schools", "We help students make realistic connections between education and the world beyond", "Choose a major from over 120 undergraduate programs and, later, from over 60 graduate programs", "Study and work on projects with our award-winning faculty", "Learn job-seeking skills and build your resume", "Take advantage of our honors programs in the arts; sciences; or social sciences", and "We are one of only two colleges in the U.S. to offer study abroad opportunities to freshmen". The least positive votes got this slogan: "We have new sports fields and the latest exercise equipment".

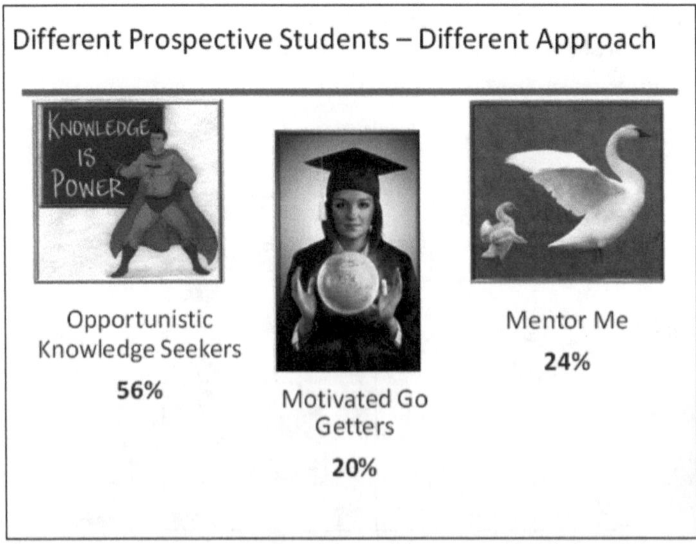

With Permissions of Howard Moskowitz And iNovum

The end result of the study was that there were three major ViewPoint mind sets named as follows: (a) Opportunity knowledge seekers, (b) Motivated go getters, and (c) Mentor me. The six figures to follow, two figures per mind type, present the poll output and type characteristics. The poll output gives the slogans and the related numerical values telling about the slogan's relevance for the given mind type. The type characteristics specify the following mayor attributes: what turns the prospect on, what turns him/her off, and, possibly, what is irrelevant. These characteristics are to be kept in mind during the communication with a prospective student once his/her mind set has been determined. As explained before, the mind set is determined after at least three multiple choice questions are placed before the incoming prospective student.

Opportunistic Knowledge Seekers (Seg1)
Interested in variety of academic programs & flexible schedules.
Not concerned with athletic amenities on campus.

1) Would you attend THIS college? <-- Definitely NO Definitely YES --> 1 2 3 4 5 6 7 8 9 Sorted by Total Sample : Highlighted >+9 winners & <-9 losers	Seg1 of 3 Knowledge
Base Size	175
Constant	49
Choose a major from over 120 undergraduate programs and, later, from over 60 graduate programs	[redacted]
Flexible schedules: You can attend classes part time or full time, day or night, and even on weekends	[redacted]
Take advantage of our honors programs in the arts; sciences; or social sciences	[redacted]
We offer the best value for your money-and a great education!	[redacted]
We help students make realistic connections between education and the world beyond	[redacted]
Meet smart and interesting students from all over the world-on a campus located in America's most ethnically diverse county	[redacted]
Our advisors help you plan your course schedule and explain degree requirements, campus services, and all the opportunities the college offers	[redacted]
Students report in top college guides that they feel "safe" and "comfortable" on our campus	-3
We offer a Bachelor of Business Administration with three areas of concentration and a unique minor in business and the liberal arts	-4
We have new sports fields and the latest exercise equipment	-4

With Permissions of Howard Moskowitz And iNovum

The characteristics of the first ViewPoint nicknamed "Opportunistic Knowledge Seekers" are that they were interested in a variety of academic programs and flexible schedules and were not concerned with athletic amenities on campus. The most appreciated slogans for this group were: "Choose a major from over 120 undergraduate programs and, later, from over 60 graduate programs.", "Flexible schedules: You can attend classes part time or full time, day or night, and even on weekends.", "Take advantage of our honors programs in the arts; sciences; or social sciences.", "We offer the best value for your money-and a great education!", and "Meet smart and interesting students from all over the world-on a campus located in America's most ethnically diverse county". The least appreciated slogans were: "Students report in top college guides that they feel 'safe' and 'comfortable' on our campus.", "We offer a Bachelor of Business Administration with three areas of concentration and a unique minor in business and the liberal arts.", and "We have new sports fields and the latest exercise equipment.".

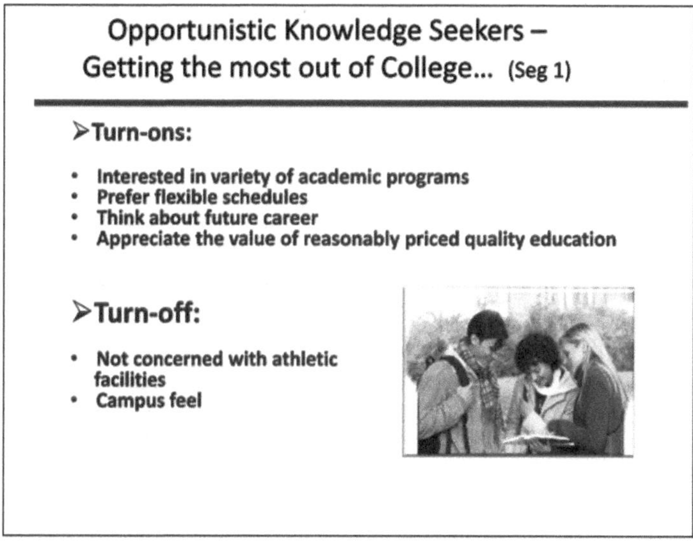

With Permissions of Howard Moskowitz And iNovum

After the ViewPoint #1 of a prospective applicant is determined using a small questionnaire, the suitable and attracting topics which college hosts should discuss with him/her are displayed in the figure corresponding to this paragraph under the subtitle "Turn-ons", and the ones that should be avoided are under "Turn-off". The Mind Genomics study had concluded that a person with this mind set is interested in variety of academic programs, prefers flexible schedules, thinks about future schedules, and appreciates the value of reasonably priced quality education. What (s)he is not interested in are sport and overall life in the campus.

Motivated Go Getters (Seg2) – Interested in continuing education at top grad schools & award-winning faculty. Not concerned with career & alumni connections.

1) Would you attend THIS college? <-- Definitely NO Definitely YES --> 1 2 3 4 5 6 7 8 9 Sorted by Total Sample : Highlighted >+9 winners & <-9 losers	Seg2 of 3 Ambitions
Base Size	61
Constant	63
Our students are accepted into the nation's top graduate schools	17
Students report in top college guides that they feel "safe" and "comfortable" on our campus	17
Take advantage of our honors programs in the arts; sciences; or social sciences	14
Study and work on projects with our award-winning faculty	12
Receive a solid education in the liberal arts and sciences	10
Be part of a college where students become leaders who can change the world	10
We're one of only two colleges in the U.S. to offer study abroad opportunities to freshmen	10
At Freshman Orientation you'll meet faculty, staff, other new students and upper classmates and learn what college life is all about	-7
Immerse yourself in our vibrant cultural life on campus with museums, concerts, theatre, dance performances, famous visiting writers.	-9
We link knowledge and skills to jobs and careers	-9
Meet and network with alumni leaders in various fields	-12

With Permissions of Howard Moskowitz And iNovum

The characteristics of the second ViewPoint nicknamed "Motivated Go Getters" are that they were interested in continuing education at top graduate schools & award-winning faculties. Also, they were not concerned with career & alumni connections. The most appreciated slogans for this group were: "Our students are accepted into the nation's top graduate schools.", "Students report in top college guides that they feel "safe" and "comfortable" on our campus.", "Take advantage of our honors programs in the arts; sciences; or social sciences.", "Study and work on projects with our award-winning faculty", and "Receive a solid education in the liberal arts and sciences". The least appreciated slogans were: "Meet and network with alumni leaders in various fields. We link knowledge and skills to jobs and careers.", and "Immerse yourself in our vibrant cultural life on campus with museums, concerts, theatre, dance performances, famous visiting writers.".

Motivated Go Getters (Seg 2)

➢Turn-ons:

- **Interested in continuing education on graduate level**
- **Want to work on projects with best faculty**
- **Care about feeling safe on campus**
- **Would like to become leaders who change the world**

➢Turn-offs:

- **Meeting alumni**
- **Linking education & career**

With Permissions of Howard Moskowitz And iNovum

After the ViewPoint #2 of a prospective applicant is determined using a small questionnaire, the suitable and attracting topics which college hosts should discuss with him/her are displayed in the figure corresponding to this paragraph under the subtitle "Turn-ons", and the ones that should be avoided are under "Turn-off". The Mind Genomics study had concluded that a person with this mind set is interested in continuing education on graduate level, wants to work on projects with best faculty, cares about feeling safe on the campus, and would like to become leader who changes the world. What (s) he is not interested in are meeting alumni and linking education and career.

Mentor Me (Seg3)

Interested in Freshmen programs to adjust to college life.
Not concerned with honors programs & other academic offerings yet.

1) Would you attend THIS college? <-- Definitely NO Definitely YES --> 1 2 3 4 5 6 7 8 9 Sorted by Total Sample : Highlighted >+9 winners & <-9 losers	Seg3 of 3 Help me to start
Base Size	74
Constant	47
Our college is minutes away from Manhattan-you can even see the skyline from our Quad	19
We're one of only two colleges in the U.S. to offer study abroad opportunities to freshmen	15
Freshman Year Initiative: Become part of a learning community of freshmen who form friendships while taking courses together	15
Our many scholarship and financial aid opportunities make it easier for you to focus on your new college life	14
Enjoy a campus with lots of cafes, dining areas and places to meet	14
We offer internships in many fields, from the media to finance, as well as job-placement services	14
Freshman Year Initiative helps you make a smooth transition from high school to college	13
Learn job-seeking skills and build your resume	13
Students report in top college guides that they feel "safe" and "comfortable" on our campus	12
Employers know that our graduates are well prepared	12
We help students make realistic connections between education and the world beyond	11
Receive a solid education in the liberal arts and sciences	-6
Be part of a college where students become leaders who can change the world	-6
Take advantage of our honors programs in the arts; sciences; or social sciences	-9

With Permissions of Howard Moskowitz And iNovum

The characteristics of the third ViewPoint nicknamed “Mentor Me” were that they were interested in freshmen programs to adjust to college life, and were not concerned with honors programs and other academic offerings yet. The most appreciated slogans for this group were: “Our college is minutes away from Manhattan-you can even see the skyline from our Quad.”, “We’re one of only two colleges in the U.S. to offer study abroad opportunities to freshmen.”, “Freshman Year Initiative: Become part of a learning community of freshmen who form friendships while taking courses together.”, “Our many scholarship and financial aid opportunities make it easier for you to focus on your new college life”, and “Enjoy a campus with lots of cafes, dining areas and places to meet”. The least appreciated slogans were: “Take advantage of our honors programs in the arts; sciences; or social sciences.”, “Be part of a college where students become leaders who can change the world.”, and “Receive a solid education in the liberal arts and sciences.”

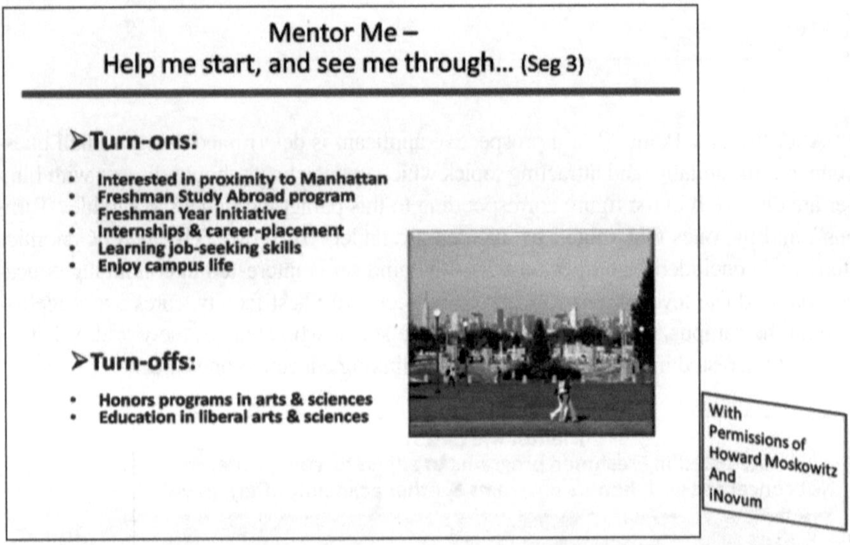

After the ViewPoint #3 of a prospective applicant is determined using a small questionnaire, the suitable and attracting topics which college hosts should discuss with him/her are displayed in the figure corresponding to this paragraph under the subtitle “Turn-ons”, and the ones that should be avoided are under “Turn-off”. The Mind Genomics study had concluded that a person with this mind set is interested in proximity to Manhattan, Freshman Study Abroad program, Freshman Year Initiative, Internships and career-placement, Learning job-seeking skills, and that (s)he enjoys campus life. What (s)he is not interested in are honors programs in arts & sciences and education in liberal arts & sciences.

Emotions: Comfortable is the strongest, followed by eager

How do you feel when you read these messages?	Eager	Uncertain	Comfortable	Intimidated	Curious
Eager					
We offer exciting, highly regarded programs in music, theatre, and the visual and performing arts	12	3	9	0	2
Our students are accepted into the nation's top graduate schools	11	3	8	1	5
Study and work on projects with our award-winning faculty	11	4	6	2	4
Uncertain					
We offer a Bachelor of Business Administration with three areas of concentration and a unique minor in business and the liberal arts	6	10	8	2	2
We graduate more future teachers than any school in the region	4	8	8	1	4
Some of our best faculty teach freshman classes	2	7	9	5	4
Comfortable					
Students report in top college guides that they feel "safe" and "comfortable" on our campus	4	4	15	1	3
We have new sports fields and the latest exercise equipment	6	6	12	0	2
Meet smart and interesting students from all over the world-on a campus located in America's most ethnically diverse county	4	4	12	0	5
Intimidated					
Enjoy a campus with lots of cafes, dining areas and places to meet	8	2	10	5	2
Some of our best faculty teach freshman classes	2	7	9	5	4
We have over 100 clubs and teams as well as Division II sports	3	5	9	5	5
Curious					
We offer the best value for your money-and a great education!	5	4	8	3	9
Freshman Year Initiative helps you make a smooth transition from high school to college	6	0	10	2	9
Our many scholarship and financial aid opportunities make it easier for you to focus on your new college life	6	4	5	3	8

With Permissions of Howard Moskowitz And iNovum

The previous six figures are related to the first question on the vignettes. The answer to this question is used to create mind types. The answer to the second question is used to figure out the most suitable slogans for both: the broadband advertising purposes and for approaching a specific individual. For example, for each specific slogan in the group where eagerness is stressed, it is shown to how many pollees eagerness is the most important; for the slogan "We offer exciting, highly regarded programs in music, theatre, and the visual and performing arts", eagerness was the answer of 12 pollees. For the slogan "Our students are accepted into the nation's top graduate schools", eagerness was the answer of 11 pollees.

The ViewPoint Analyzer

Admissions Rep asks a prospective student a few qualifying questions at the beginning of the conversation to establish his/her ViewPoint/MindSet

To best help you achieve your online educational goals with us, please rate how important these six factors are to you;

	Not at all	Unsure	Definitely
College is close to home	○	○	○
Excellent programs to prepare and earn admittance to graduate schools	○	○	○
Extensive libraries	○	○	○
Programs have a career-focused curriculum built on the latest industry trends	○	○	○
A well maintained, beautiful campus	○	○	○
Academic Advisors available when you need support	○	○	○

Submit

With Permissions of Howard Moskowitz And iNovum

With the help of the Mind Genomics' software module ViewPoint Analyzer, a small questionnaire for determining mind types was created and was given to every prospective applicant. The questions were related to convenience, quality, and the market acceptance. A customer was supposed to read six factors relevant to the college offers and rate how important the factors were to him/her. Rating was supposed to be expressed with (a) "Not at all interested", (b) "Unsure", and (c) "Definitely interested". Those six given factors were: "College is close to home", "Excellent programs to prepare and earn admittance to graduate school", "Extensive libraries", "Programs have career-focused curriculum built on the latest industry trends", "A well maintained, beautiful campus", and "Academic advisors available when you need support". The answers revealed the mind type of the person who opted to take the value exchange offer and whose mind type was to be classified.

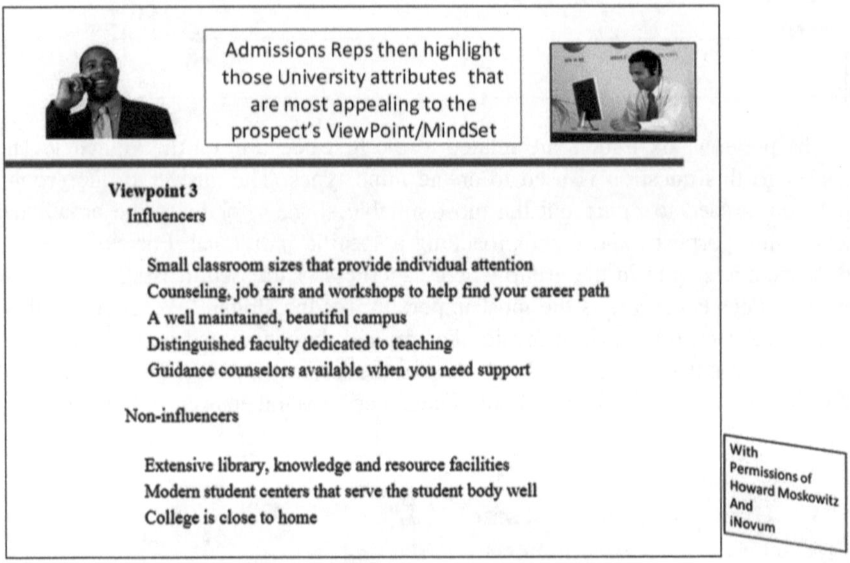

After the mind type of a customer was determined, a college officer gets the exact list of topics and slogans that should be discussed with the customer and which ones should be avoided. The related figure contains issues of relevance for the ViewPoint #3. For the case of influencers, these were: (a) "Small classroom sizes that provide individual attention", "Counseling, job fairs and workshops to help find your career path", "A well maintained, beautiful campus", "Distinguished faculty dedicated to teaching", and "Guidance counselors available when you need support". For the case of non-influencers, these were: (b) "Extensive library, knowledge and resource facilities", "Modern student centers that serve the student body well", and "College is close to home".

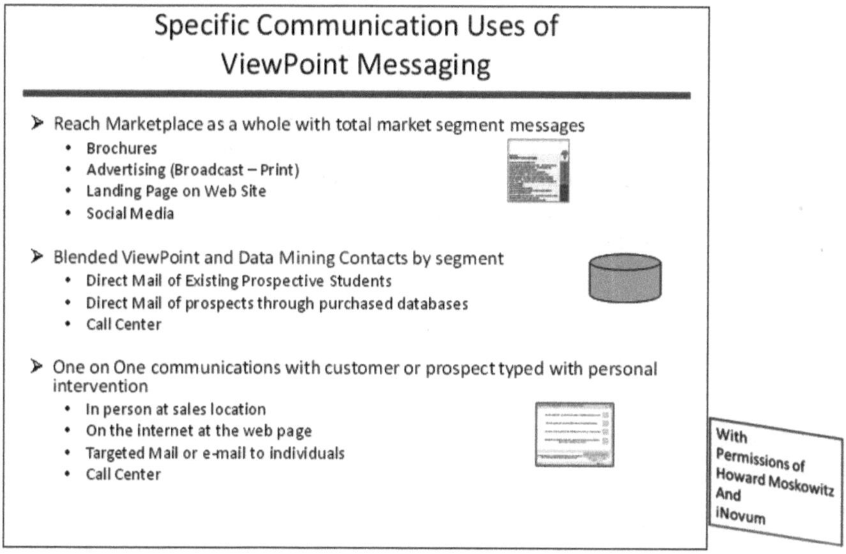

The figure related to this paragraph sheds light on specific communication uses of ViewPoint messaging. It encompasses three different domains: (a) Reaching the marketplace as a whole with total market segment messages, (b) Blending ViewPoint and Data Mining contacts by segment, and (c) One on One communicating with a customer or a prospect typed with personal intervention. Issues of relevance for the first domain are: brochures, advertising (broadcast—print), landing page on web site, and social media. Issues of relevance for the second domain are: direct mail of existing prospective students, direct mail of prospects through purchased databases, and call center. Issues of relevance for the third domain are: in person at sales location, on the internet at the web page, targeted mail or e-mail to individuals, and call center.

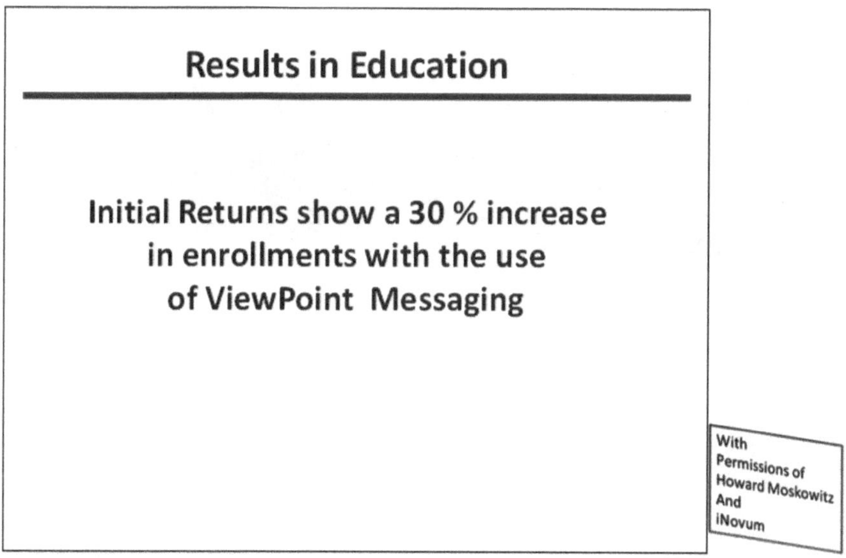

In conclusion, as indicated before, initial returns showed a 30 % increase in enrollments with the use of ViewPoint messaging. Note that the 30 % increase in enrollment may mean a considerably higher increase in profit, due to the fact that some of the costly resources used in the education process could be shared. Namely, increased enrollment helps amortize better those costly shared resources. Consequently, Mind Genomics helped with the most vital aspects of profit making. It either increased the profit or helped turn a loss into a profit. This case study was chosen to help understand better this important set of issues.

3.4 Case #4—Perfect Healthy Breakfast Cereal

The following example is not trying to woo customers to any existing product or service. It shows how Mind Genomics can help in deciding how a new product should be developed in order to be successful on the market. The goal was to determine what characteristics (ingredients, sensory specifics, price range, and brand) a "Healthy Breakfast Cereal" should have in order to be perceived as perfect by customers. Also, the goal was to discover what marketing phrases should be used to attract customers to buy newly created cereal. The analysis showed what segments exist and that there are differences between genders and income groups. It also indicated that healthy trends are continually changing and that testing should be repeated on a reasonably frequent basis.

With Permissions of Howard Moskowitz And iNovum

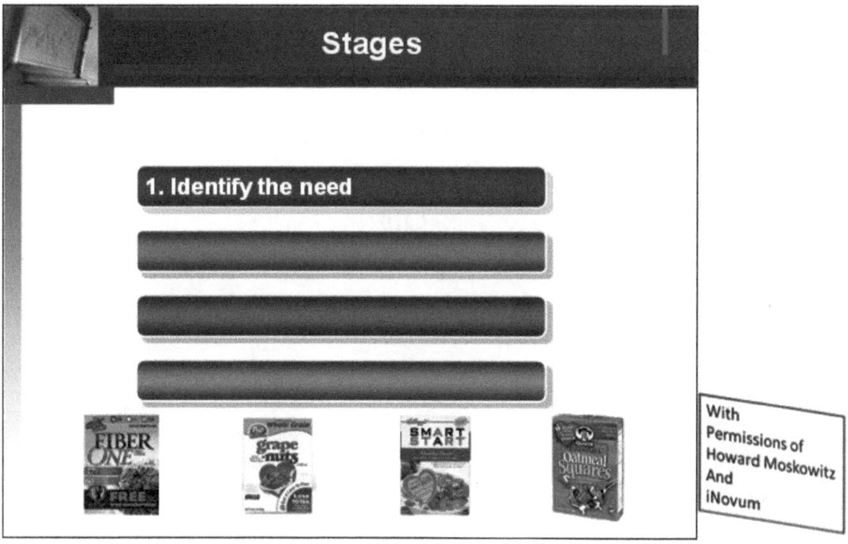

The case study shows that the Mind Genomics phase 1, if used alone (without phase 2), could also create important value in a wide plethora of applications. In this case study it helps create a new product, for which a clear need was noted: a new healthy breakfast cereal. The need arose from the understanding of the essence of the following facts (indicated in the figure corresponding to this paragraph). The first (unfortunate) fact is that more than 60 % of Americans are obese, so for all of them there is a necessity to find something healthy to eat in the morning that will provide some schemes for weight losing. Such schemes should include a high protein and low carbohydrate diet, which is known to be effective for weight loss. The study tries to define what the important aspects of a healthy breakfast cereal are and whether the perception that soy products are healthy is correct. All the above is relative to the stage one of the process. This paragraph is related to the next figure, too.

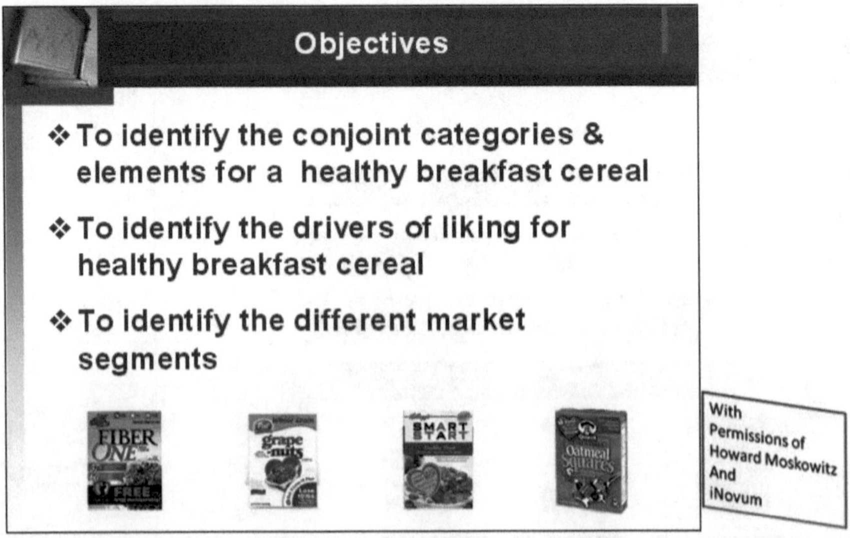

Knowing the facts related to motivation for creation of a new product, one has to set appropriate objectives. In the case study under consideration, these objectives are: (a) identifying the conjoint (those in synergy with each other) categories & elements for a healthy breakfast cereal (silos and slogans), (b) identifying the drivers of liking for healthy breakfast cereal, and (c) identifying the different market segments. Having clear objectives is probably the most important ingredient for setting the stage one properly.

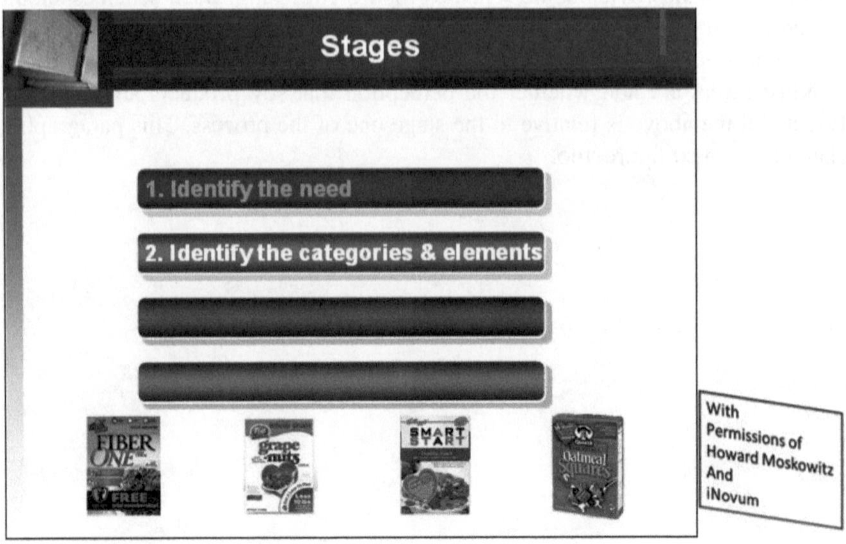

This paragraph is related to two figures—this one and the next one. In the stage two of this process one has to define categories and elements (silos and slogans). The study includes the following processes of relevance for setting the stage two: the ideas, the setup, and the test. Correctly defining basic ideas about the product is very helpful in identifying categories of interest and choosing the correct six silos. The setup consists of defining the elements for all the silos and of consolidating the elements into as good slogans as possible. The test has two phases: (a) vignettes preparation and trial testing and (b) launching and monitoring the test.

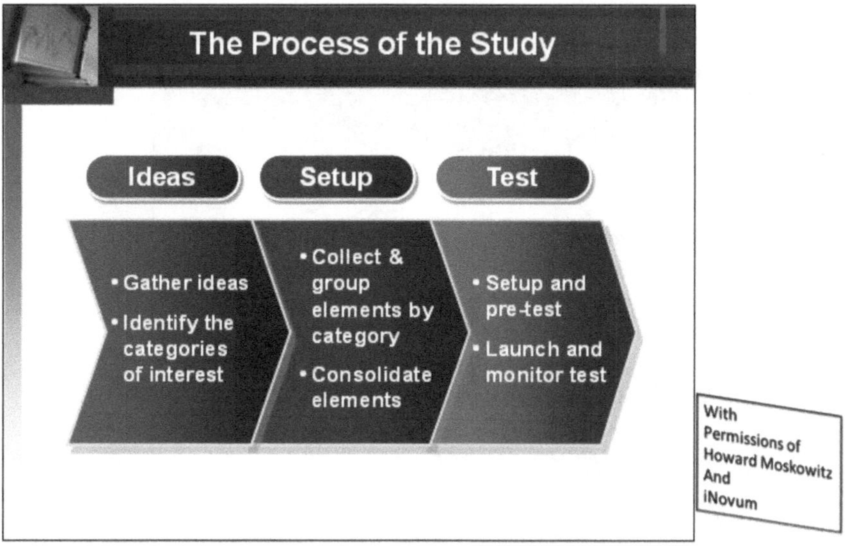

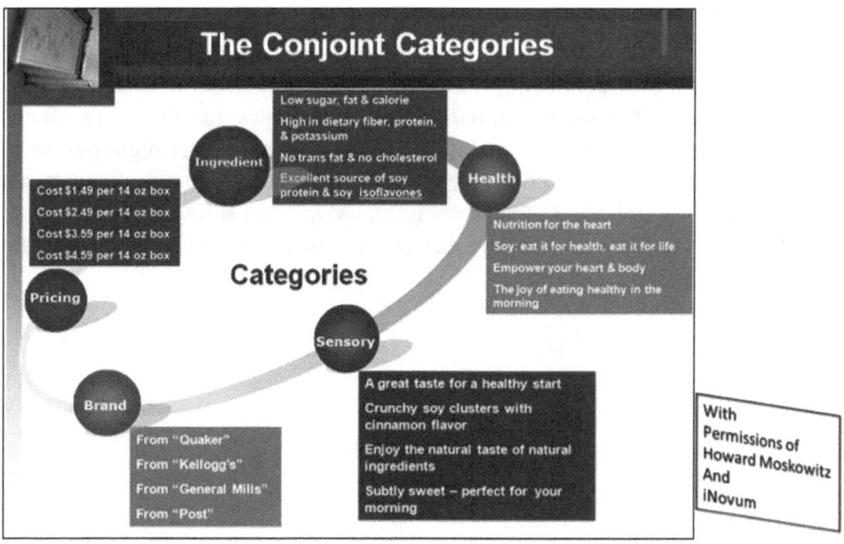

The defined five conjoint categories were: (a) Ingredients: “Low sugar, low fat & low calorie”, “High in dietary fiber, protein, & potassium”, “No trans fat & no cholesterol”, and “Excellent source of soy protein & soy isoflavones”; (b) Health: “Nutrition for the heart”, “Soy: eat it for health, eat it for life”, “Empower your heart & body”, and “The joy of eating healthy in the morning”; (c) Sensory: “A great taste for a healthy start”, “Crunchy soy clusters with cinnamon flavor”, “Enjoy the natural taste of natural ingredients”, “Subtly sweet—perfect for your morning”; (d) Brand: “From ‘Quaker’”, “From ‘Kellogg’s’”, “From ‘General Mills’”, “From ‘Post’”; and (e) Price: Cost $1.49, $2.49, $3.59, and $4.59 per 14 oz box.

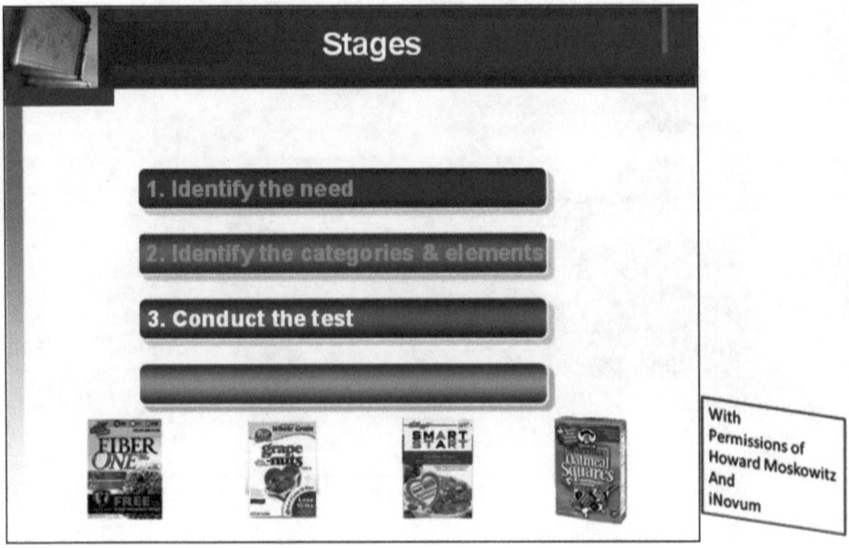

This paragraph is again related to two figures—this one and the next one. The stage three consists of conducting the test (the polling). One first has to initiate the test (to prepare it properly). Initiating the test includes: (a) Setting up the test when its pre-testing has to prove that it is user friendly and fast enough (pollees do not like long quizzes); (b) Recruiting enough pollees that is done via Internet; and (c) Launching & monitoring the test using IdeaMap®.Net software, which allows instant feedback, adjustments for number of participants, and can be completed in 24 h for all the participants.

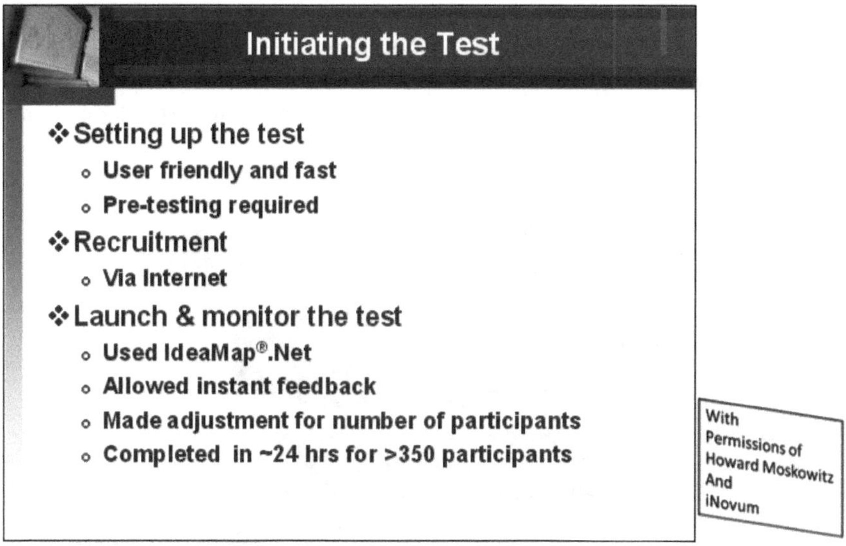

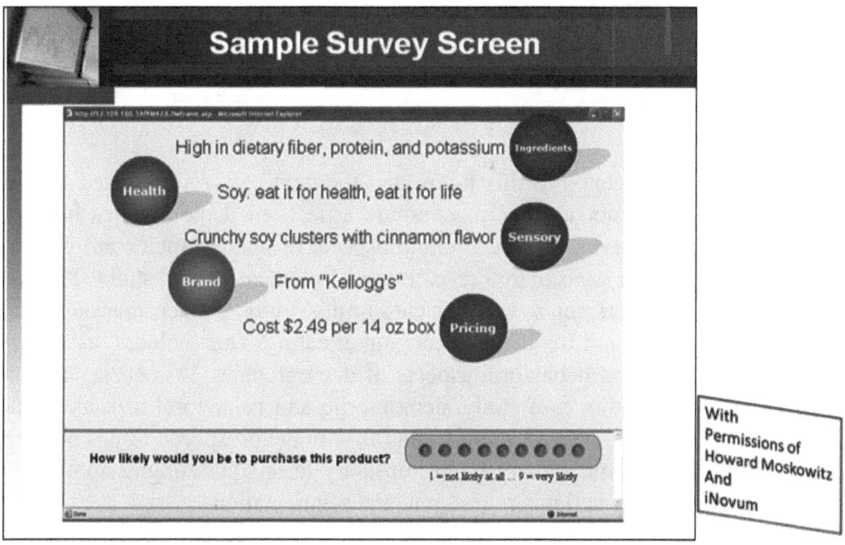

The survey screen (the vignette) should cover the five mentioned conjoint categories. One example for the conjoint category one (Ingredients) is "high dietary fiber, protein, and potassium; for the conjoint category two (Health) is "Soy, eat it for health, eat it for life"; for the conjoint category three (Sensory) is "Crunchy soy clusters with cinnamon flavor"; for the conjoint category four (Brand) is "From 'Kellogg's'" and for the conjoint category five (Pricing) the example is

"Cost $2.49 per 14 oz box". Note that in this polling each vignette has five and not four slogans, which is the consequence of the reality that five conjoint categories are defined. The stated question on all vignettes was "How likely would you be to purchase this product?" The answer was given by rating from 1 (not likely at all) to 9 (very likely).

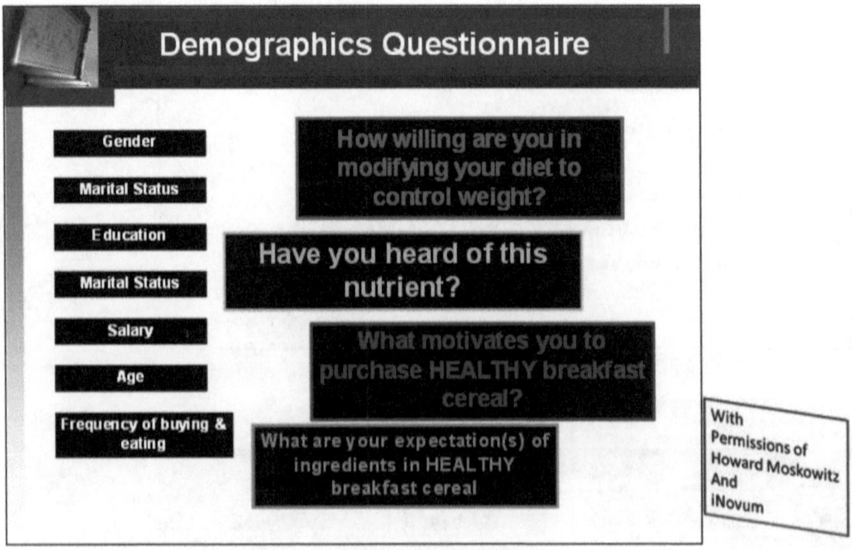

This paragraph is related to this figure and the next one. As indicated before, Mind Genomics does not use socio-economic aspects of demographics in mind typing (phase 2); however, socio-economic aspects of demographics are used in micro science (phase 1) and therefore is a crucial element of this study. The following demographics elements were selected in this study: gender, marital status, education, salary, age, and frequency of buying & eating. This includes socio-economic, demomorfic, and behavioral aspects of demographics. Of course, had the phase 2 been used in this case study, demomorfic and behavioral aspects would have been used to create conversational patterns with the prospects. Issues of interest for the related statistical analysis are: ordinary least square regressions (individual models), dummy variable regression, and segmentation.

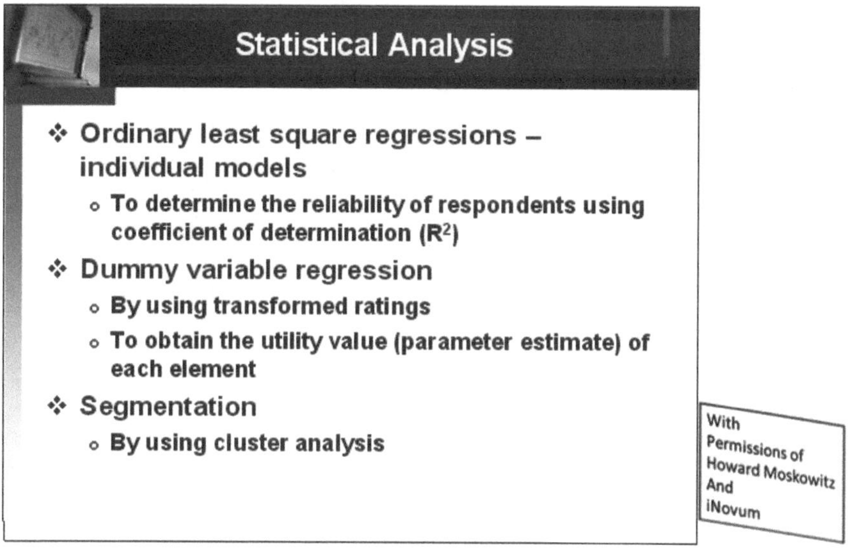

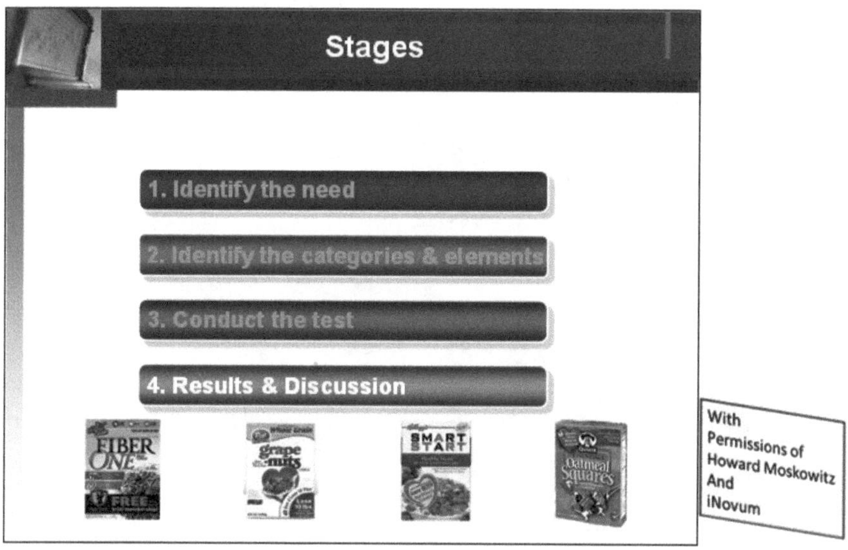

The stage four includes the analysis of the results. The study is more trustable if the level of respondents' consistency is higher, i.e. if the following results pile up around a few stacking points. The graph on the next figure, with the coefficient of determination on the X axes and the number of participants on the Y axes tells about the respondents' consistency. In the case of this study, one can see on the chart that the number of respondents that have coefficient of determination close to one is high. This shows that the results of the study are highly trustable due to the high level of the respondents' consistency.

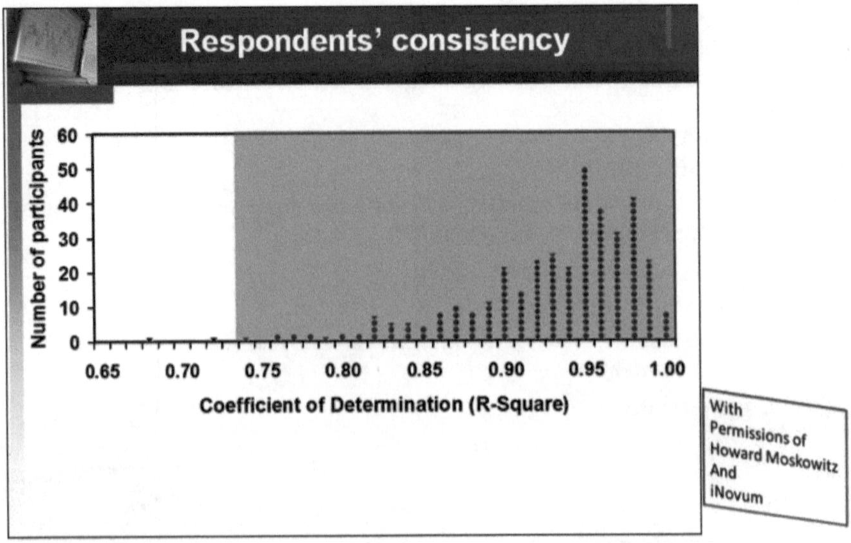

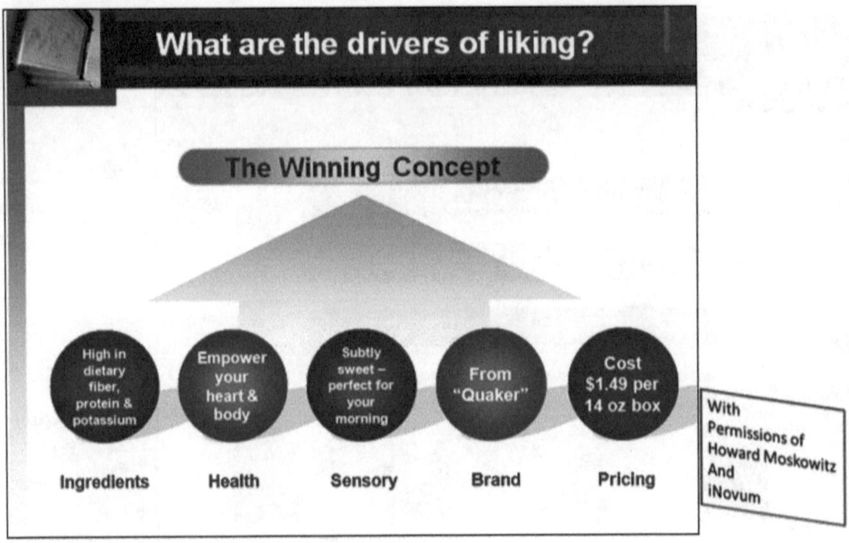

The figure corresponding to this paragraph depict the winning slogans for the five conjoint categories; the ones that should be used as the basis, both for the characteristics of the new products and for the marketing campaign that should follow products' arrival on the market. In the first category (Ingredients), they should include phrases like "high dietary fiber, protein, and potassium"; in the category two (Health), "empower your heart & body"; in the category three (Sensory), "subtly sweet—perfect for your morning"; in the category four (Brand), "from 'Quaker'", and in the category five (Pricing), the example is "cost $1.49 per 14 oz box".

The figure corresponding to this paragraph depict the slogans for the five conjoint categories, which should not be used as the basis, either for the characteristics of the new products, or for the marketing campaign that follows products' arrival on the market. In the first category (Ingredients), one should exclude phrases like "excellent source of soy protein & soy isoflavones"; in the category two (Health), "soy—eat it for health, eat it for life"; for the category three (Sensory), is "a great taste for a healthy start"; for the category four (Brand), is "from 'Post'", and for the category five (Pricing), the example is "cost $4.59 per 14 oz box".

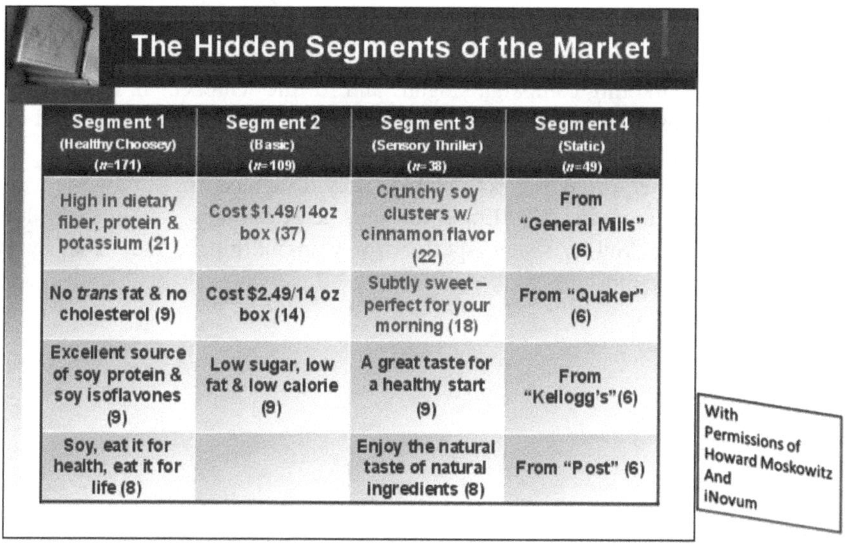

Each market includes hidden segments that may have a non trivial impact. The four hidden segments of relevance are: Health (segment 1), Basic (segment 2) Sensory (segment 3) and Static (segment 4). Segment 1 includes issues like “High in dietary fiber, protein & potassium” and “Excellent source of soy protein & soy isoflavones”; Segment 2 includes issues like “Cost $1.49/14 oz box” and “Low sugar, low fat & low calorie”; Segment 3 includes issues like “Crunchy soy clusters w/cinnamon flavor” and “Subtly sweet—perfect for your morning”; Segment 4 is equally spread over four renown cereal producers.

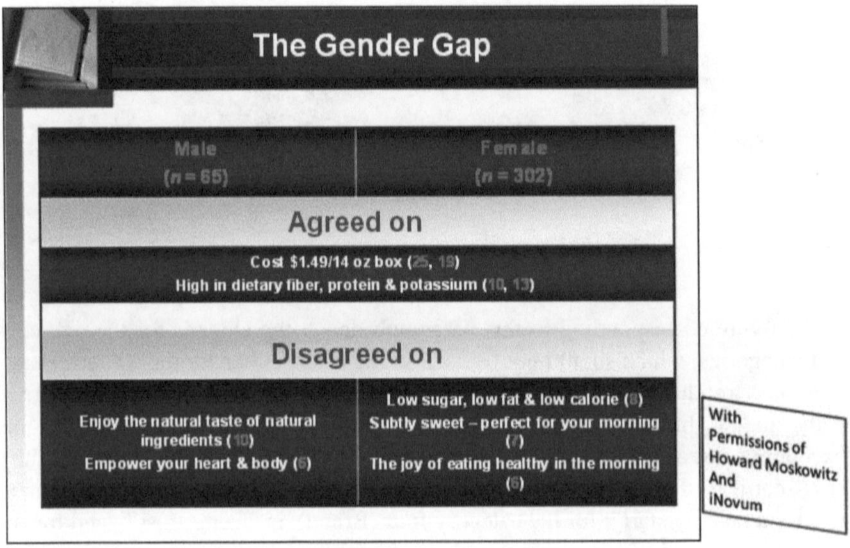

Issues related to the genders gap are relevant for the deep understanding of the results of the phase 1 and, once the new product is generated, for the proper wording of the marketing campaign slogans and for the sentences in the communications with prospects. In this study the number of male pollees was 65 and the number of female pollees was 302. They agreed on “Cost $1.49/14 oz box” and “High in dietary fiber, protein & potassium”. They disagreed on: “Enjoy the natural taste of natural ingredients”, “Empower your heart & body”, “Low sugar, low fat & low calorie”, “Subtly sweet—perfect for your morning”, and “The joy of eating healthy in the morning”.

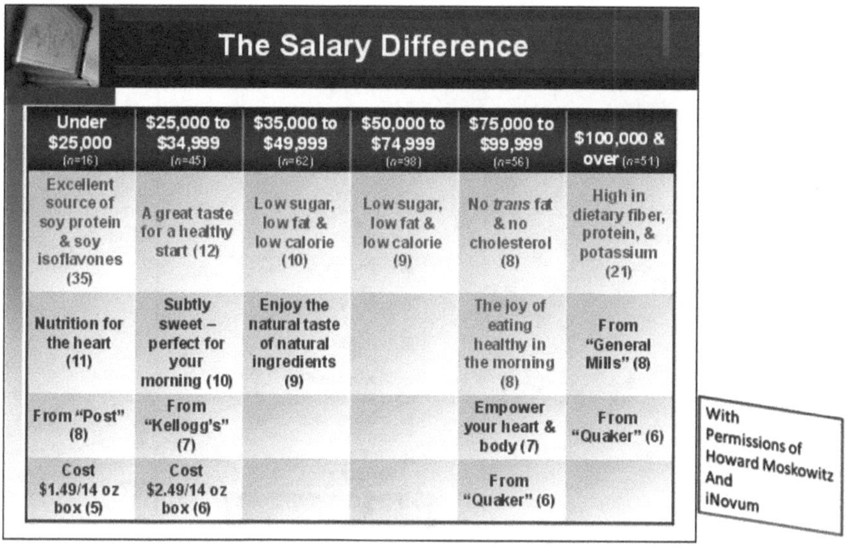

The Salary Difference

Under $25,000 (n=16)	$25,000 to $34,999 (n=45)	$35,000 to $49,999 (n=62)	$50,000 to $74,999 (n=98)	$75,000 to $99,999 (n=56)	$100,000 & over (n=51)
Excellent source of soy protein & soy isoflavones (35)	A great taste for a healthy start (12)	Low sugar, low fat & low calorie (10)	Low sugar, low fat & low calorie (9)	No *trans* fat & no cholesterol (8)	High in dietary fiber, protein, & potassium (21)
Nutrition for the heart (11)	Subtly sweet – perfect for your morning (10)	Enjoy the natural taste of natural ingredients (9)		The joy of eating healthy in the morning (8)	From "General Mills" (8)
From "Post" (8)	From "Kellogg's" (7)			Empower your heart & body (7)	From "Quaker" (6)
Cost $1.49/14 oz box (5)	Cost $2.49/14 oz box (6)			From "Quaker" (6)	

The figure corresponding to this paragraph presents the winning slogans for different salary groups. Issues of importance for those of under 25.000 US$ were: "Excellent source of soy protein & soy isoflavones", "Nutrition for the heart", "From 'Post'", and "Cost $1.49/14 oz box"; for those of over 25.000 and under 35,000 US$ were: "A great taste for a healthy start", "Subtly sweet—perfect for your morning", "From 'Kellogg's'", and "Cost $2.49/14 oz box"; for those of over 35.000 and under 50.000 US$ were: "Low sugar, low fat & low calorie" and "Enjoy the natural taste of natural ingredients"; for those of over 50.000 and under 75.000 US$ was: "Low sugar, low fat & low calorie"; for those of over 75.000 and under 100.000 US$ were: "No *trans* fat & no cholesterol", "The joy of eating healthy in the morning", "Empower your heart & body", and "From 'Quaker'"; and for those of over 100.000 US$ were: "High in dietary fiber, protein, & potassium", "From 'General Mills'", and "From 'Quaker'".

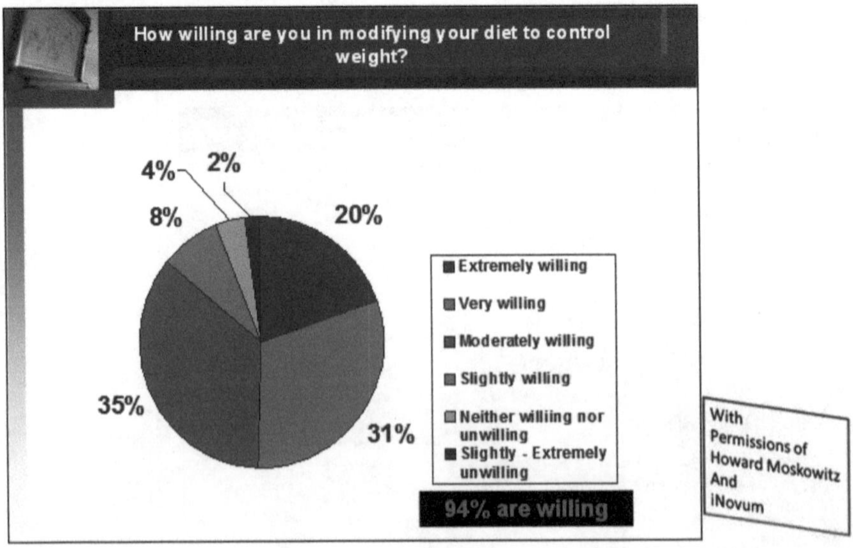

The next important parameter for the decisions in this study is how willing the prospective customers are to modify their diet. The pie chart from the figure related to this paragraph reveals the following facts: 20 % of the respondents are extremely willing to modify their diet, 31 % of them are very willing, 35 % of them are moderately willing, 8 % of them are slightly willing, while 4 % are neither willing or unwilling, and only 2 % of the respondents are unwilling to modify their diet, Altogether 94 % of respondents are willing to modify their diet.

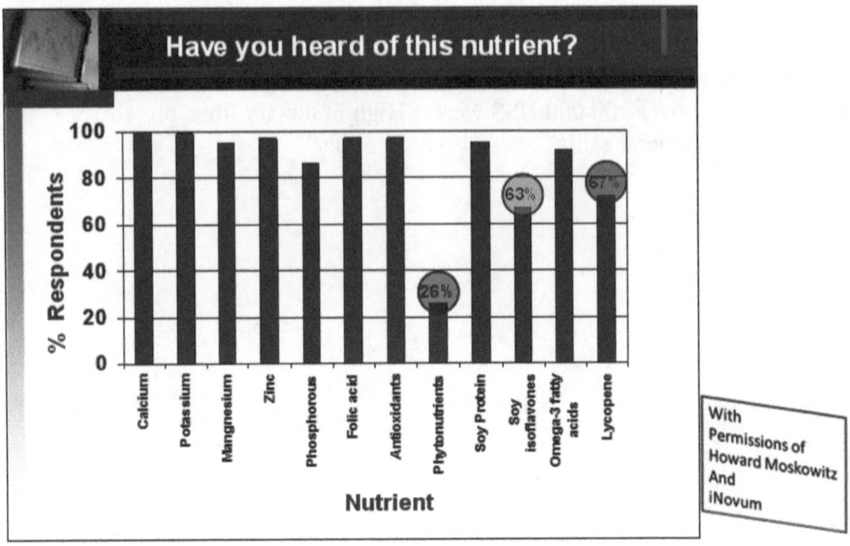

Response of the prospects depends also on their educational level, at least for the understanding of the impact of the nutrients involved in the diet. The graph in the figure related to this paragraph connects the nutrient type and the count of respondent that have heard for the nutrient. Almost all of the respondents have heard for calcium, potassium, zinc, folic acid, antioxidants, magnesium, soy protein, omega 3 fatty acids, and phosphorous. They were not very familiar with lycopane and soy isoflavones, and they were the least familiar with phytonutrients. The conclusion is that these three facts have to be given a very careful attention in the process of marketing this product.

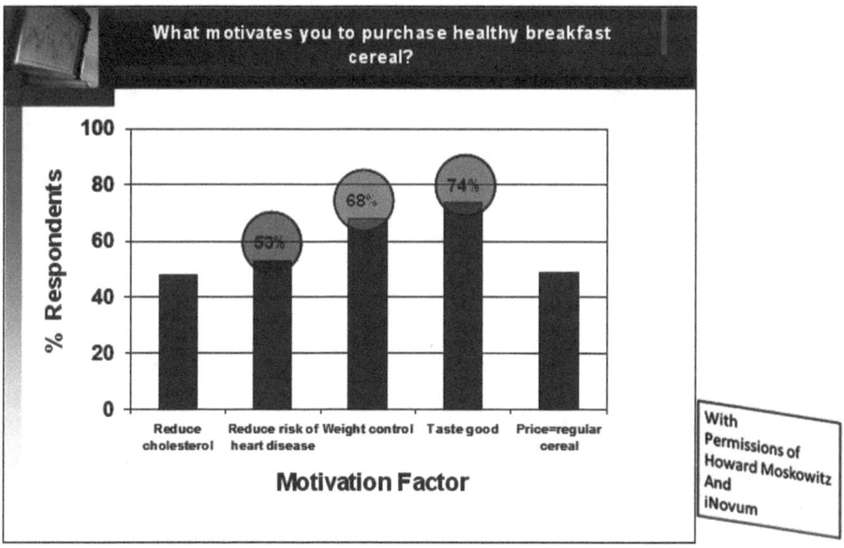

Motivation factors are also relevant for the marketing campaign. The histogram related to this paragraph connects the main motivation factors and the number of respondents. It reveals the following—the motivation factor that is the strongest one (appreciated by 74 % of the respondents) is how the cereal tastes—"tastes good". The next one (68 % of the respondents) is how eating the cereal controls the weight. The third one (53 % of the respondents) is whether eating the cereal would help in fighting heart disease—"Reduces risk of a heart disease". The fourth ranked motivation factor (48 % of the respondents) is the price compared to the regular cereal, and the fifth motivation factor (46 % of the respondents) is how the cereal reduces bad cholesterol. Those sticking out factors are to be stressed in the marketing campaign.

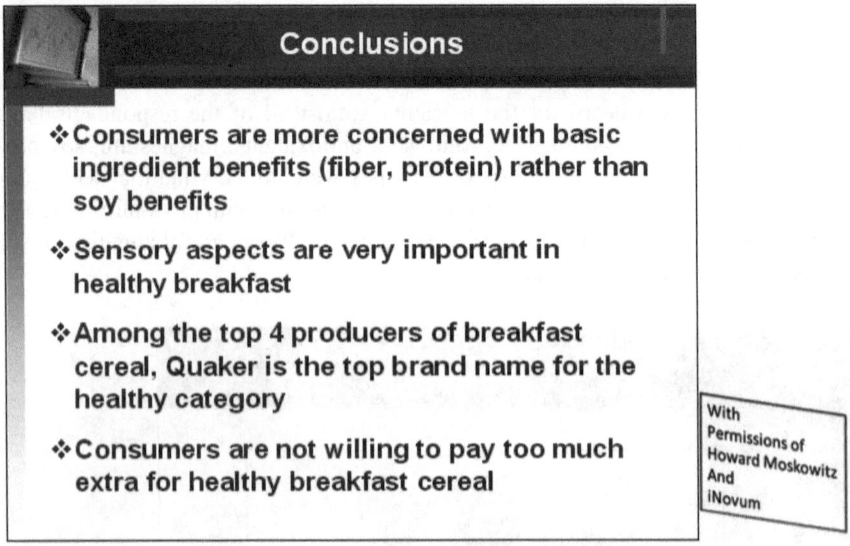

This paragraph is related to this figure and the next one, i.e. to the Conclusions: Consumers were more concerned with basic ingredient benefits (fiber, protein) rather than soy benefits and were not willing to pay too much extra for healthy breakfast cereal. They thought that sensory aspects are very important in healthy breakfast, and they were also easily led by the brand name—the most popular was Quaker. The study has also shown that gender and income differences exist in segment clusters. The soy and soy proteins did not fare well in contrast to other proteins. Also, it was clearly proved that things that are perceived as healthy continuously change, so the retesting (repeating the study) is needed from time to time.

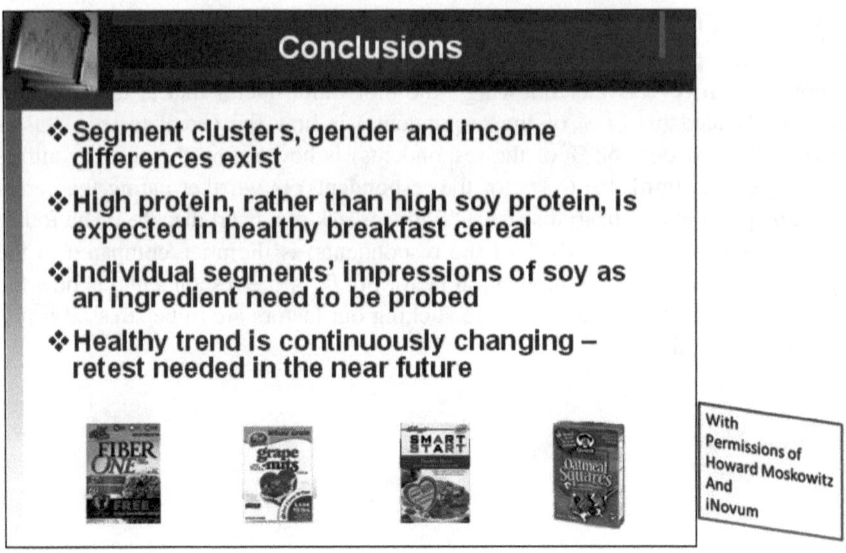

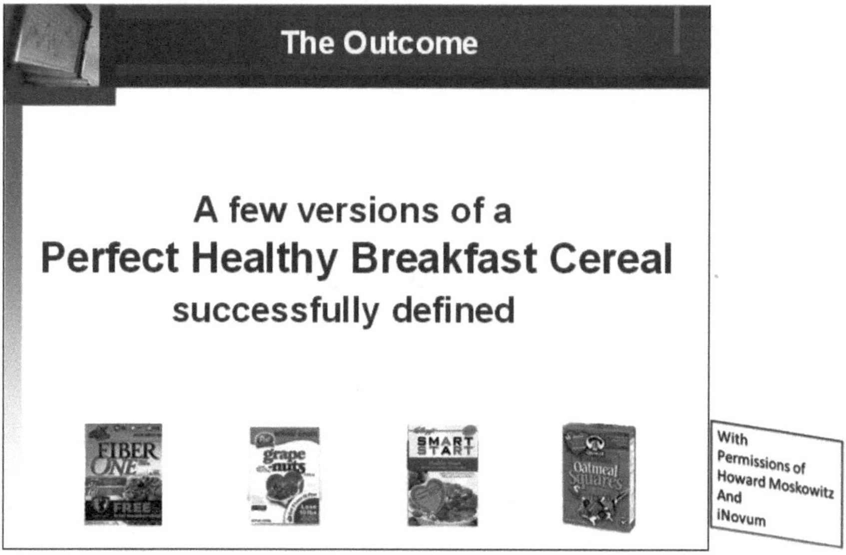

In essence, this study defined successfully not one, but a few versions of the healthy breakfast cereal, using the concept of horizontal segmentation. Actually, a side product of the study is the horizontal segmentation which, based on responses of the pollees define four different product types.

3.5 Case #5—Banking Industry

The following example is from yet another different industry—financials and banking, where Mind Genomics showed even better results. This study was the case of one small community bank in the Southwest of USA. The case study discovered four Addressable mind segments called: "self driven on-line", "high security", "collaborative on-line", and "personal touch". After the Mind Genomics science was performed, members of all segments had the same impression when approaching the bank—"The bank knows me!". The presented figures show how Mind Genomics can, by creating correct lists of "Turn-on" and "Turn-off" slogans, help bank's employees to reduce negative messaging by 57 % and to increase positive messaging by a staggering 460 % thus increasing customers' satisfaction a great deal.

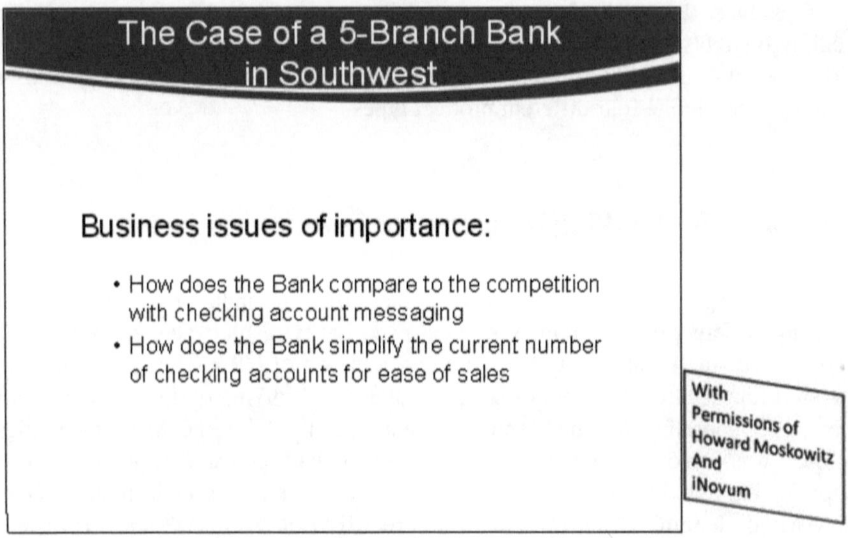

This paragraph is related to this figure and the next one. In the case of the bank under consideration, the major two issues of importance were: (a) How the Bank compares to the competition with checking account messaging and (b) How the Bank could simplify the number of different checking accounts for ease of sales. The Mind Genomics study defined two mind types (segments) of relevance for this study: (a) Price & Simplicity and (b) Relation & Service. The slogans that induced the most positive and the most negative reactions of the total polled populations are displayed in the next figure. As example they were: (a) the most positive: "Totally Free! … no minimum balance—no monthly service charge", "No monthly charges with direct deposit", and "Includes Identity Protection"; and the most negative: "$8 monthly fee", "Free with $25 automatic monthly transfer from your savings account", and "Meet 3 monthly qualifications—eliminate ATM surcharges".

Checking Account Message Rating by Total Market: Strong Positives vs Strong Negatives

	Total Sample
CONSTANT: Propensity to open a checking account now	22
Totally Free!... no minimum balance - no monthly service charge	24
No monthly charges with direct deposit	11
Includes Identity Protection	7
Open a checking account and receive a preferred rate on new savings accounts for each household member	6
Includes on-line & mobile banking & bill pay -- for the online power user	6
Neighborhood bankers you can trust	6
Earn rewards points with each use à Redeem your rewards for CA$H	6
One free surcharge ATM withdrawal per month	-6
Bounced check protection - for a $32 fee we will not bounce checks on overdrawn accounts	-7
Local strength in banking	-7
Meet 3 monthly qualifications - eliminate ATM surcharges	-9
Free with $25 automatic monthly transfer from your savings account	-14
$8 monthly fee	-22

With Permissions of Howard Moskowitz And iNovum

Segment messaging often conflicts with each other...
Segment 1: Strong Positive for price and Strong Negative for additional services

	Total Sample	Price & Simplicity	Relation & Service
CONSTANT: Propensity to open a checking account now	22	49	9
Totally Free!... no minimum balance - no monthly service charge	24	30	21
No monthly charges with minimum balance	0	8	-4
Neighborhood bankers you can trust	6	-8	13
Hassle free switching-- we move your checking account from your existing bank at no charge	5	-8	11
One free surcharge ATM withdrawal per month	-6	-9	-5
Count on us	-2	-10	1
$8 monthly fee	-22	-10	-28
Convenient surcharge free ATMs 1,200 local to where you live 37,000 Nationwide when you travel	1	-12	8
Earn rewards points with each use à Redeem your rewards for CASH	6	-12	15
Free single signer American Express Traveler Cheques	-2	-11	4
Five free stop payments per year	2	-14	10
Have it your way-- several checking accounts to choose from	-1	-14	6
Bank or pay bills anytime from where you live, work, play or travel	3	-16	11
Meet 3 monthly qualifications - eliminate ATM surcharges	-9	-12	-5
Local strength in banking	-7	-16	-2
Earn rewards points with every use à Redeem your rewards for cash, travel or merchandise	-3	-19	5
Qualify for balance interest rewards - meet 3 simple qualifiers per statement cycle	-4	-19	3
Free with $25 automatic monthly transfer from your savings account	-14	-19	-12
Free first order of checks	-2	-20	7
You receive 5,000 Rewards bonus points when you open your checking account with direct deposit	3	-21	15

With Permissions of Howard Moskowitz And iNovum

The characteristics of the first ViewPoint nicknamed "Price & Simplicity" are that they were strong positive for price and strong negative for additional services. The most appreciated slogans for this group were: "Totally Free! … no minimum balance—no monthly service charge" and "No monthly charges with minimum balance". The least appreciated were: "Bank or pay bills anytime from where you live, work, play or travel", "Meet 3 monthly qualifications—eliminate ATM surcharges", "Local strength in banking", "Earn rewards points with every use and redeem your rewards for cash, travel or merchandise", "Quality for balance interest rewards—meet 3 simple qualifiers per statement cycle", "Free with 625 automatic monthly transfer from your savings account", "Free first order of checks", and "You receive 5.000 Rewards bonus points when you open your checking account with direct deposit".

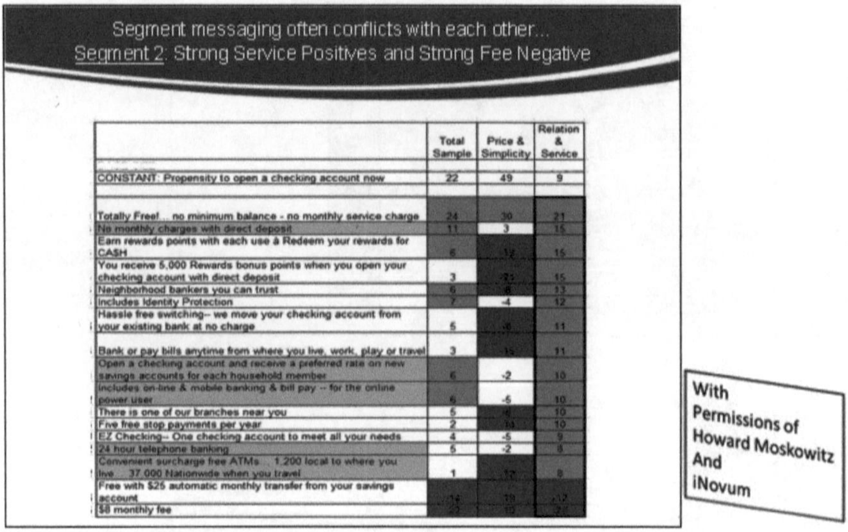

	Total Sample	Price & Simplicity	Relation & Service
CONSTANT: Propensity to open a checking account now	22	49	9
Totally Free! … no minimum balance - no monthly service charge	24	30	21
No monthly charges with direct deposit	11	3	15
Earn rewards points with each use à Redeem your rewards for CASH	6	[illegible]	15
You receive 5,000 Rewards bonus points when you open your checking account with direct deposit	3	[illegible]	15
Neighborhood bankers you can trust	6	[illegible]	13
Includes Identity Protection	7	-4	12
Hassle free switching-- we move your checking account from your existing bank at no charge	5	[illegible]	11
Bank or pay bills anytime from where you live, work, play or travel	3	[illegible]	11
Open a checking account and receive a preferred rate on new savings accounts for each household member	6	-2	10
Includes on-line & mobile banking & bill pay -- for the online power user	6	-5	10
There is one of our branches near you	5	[illegible]	10
Five free stop payments per year	2	[illegible]	10
EZ Checking-- One checking account to meet all your needs	4	-5	9
24 hour telephone banking	5	-2	8
Convenient surcharge free ATMs … 1,200 local to where you live … 37,000 Nationwide when you travel	1	[illegible]	8
Free with $25 automatic monthly transfer from your savings account	[illegible]	[illegible]	[illegible]
$8 monthly fee	[illegible]	[illegible]	[illegible]

The characteristics of the second ViewPoint nicknamed "Relation & Service" are that they were strong service positives and strong fee negative. The most appreciated slogans for this group were: "Totally Free! … no minimum balance—no monthly service charge", "No monthly charges with direct deposit", "Earn rewards points with each use and redeem your rewards for CASH", "You receive 5,000 Rewards bonus points when you open your checking account with direct deposit", "Neighborhood bankers you can trust", "Includes Identity Protection","Hassle free switching– we move your checking account from your existing bank at no charge", and "Bank or pay bills anytime from where you live, work, play or travel". The least appreciated were: "$8 monthly fee" and "Free with $25 automatic monthly transfer from your savings account".

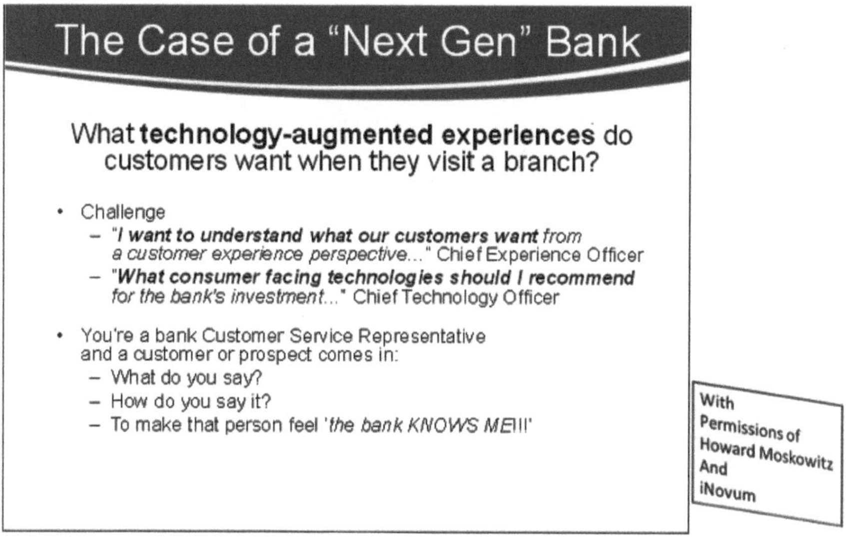

This paragraph is related to this figure and the next one. In this case study the major challenges were: (a) "How to understand what the customers want from technology-augmented experiences perspective" and (b) "What consumer facing technologies should be recommended for the bank's investment". The Mind Genomics study defined four mind types (segments) of relevance for this study dubbed: (a) Self driven online, (b) High security, (c) Collaborative Online, and (d) Personal touch. The self driven online type spends lots of time sitting at the computer. The high security type's highest priority is security, especially in online banking. The collaborative online type is not afraid of communicating with unknown persons over the Internet. The personal touch type likes to have a one to one contact with a known person, like a friendly bank's clerk.

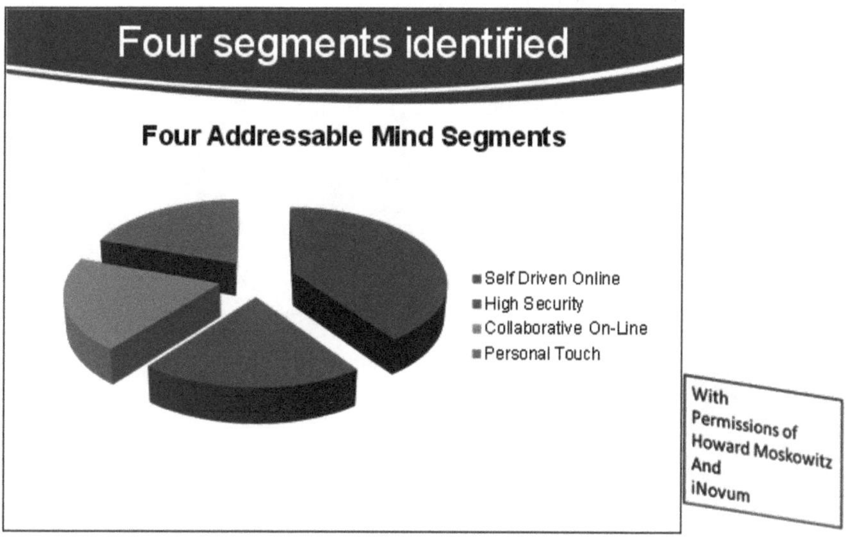

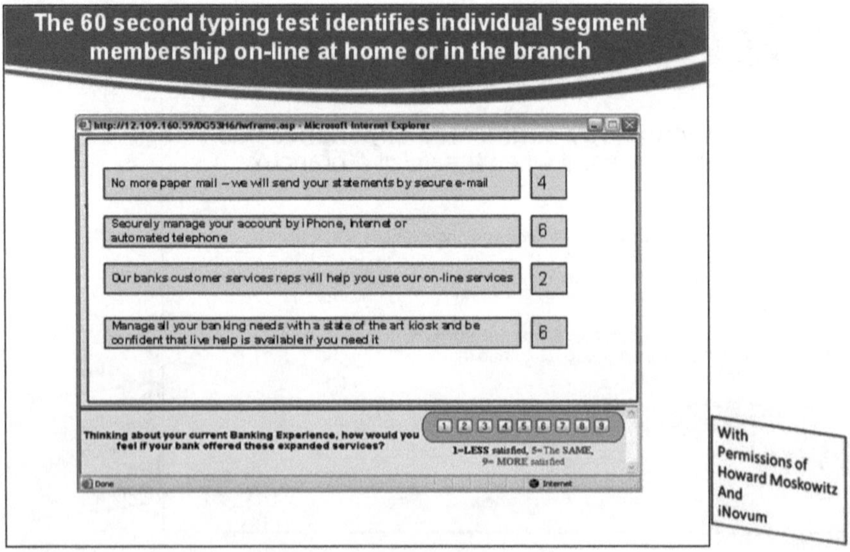

The figure relevant to this paragraph presents an example of a small questionnaire used for determining mind types of a bank's customer. The questions were related to convenience, quality, and the market acceptance. A customer was supposed to read four slogans relevant to the bank's offers and rate how (s)he would feel if the bank offered those extended services. Rating was supposed to be expressed with (a)"Less satisfied", (b) "The same", and (c) "More satisfied". The four given slogans were: "No more paper mail – we will send your statements by secure e-mail", "Securely manage your account by iPhone, Internet or automated telephone", "Our banks customer services reps will help you use our on-line services", and "Manage all your banking needs with a state of the art kiosk and be confident that live help is available if you need it".

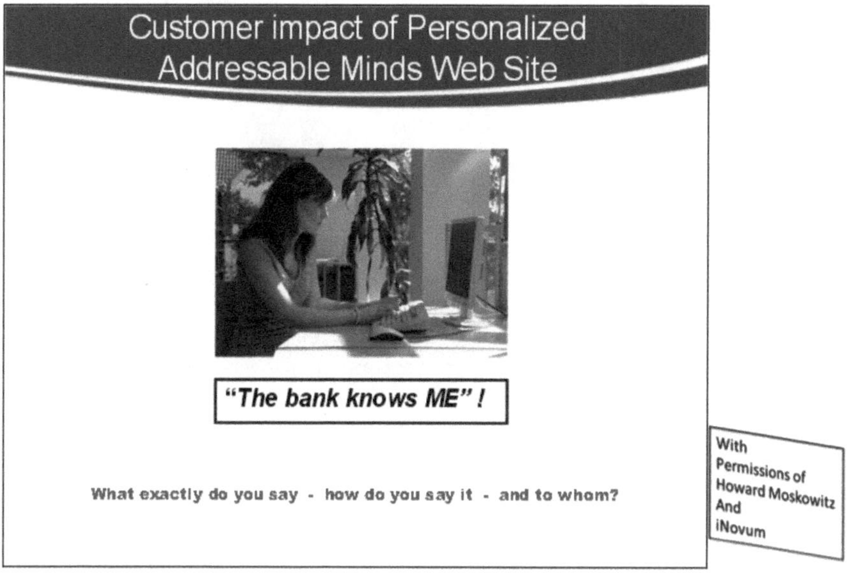

As always, the main purpose of the second phase of a Mind Genomics process is that the client gets convinced that the provider of the products or services "knows him/her". The conclusion related to these "bank" cases is that Mind Genomics can be used for a number of different purposes: (a) as an Executive decision tool, because it can tune existing product set marketing and it can develop timely new offerings to meet dynamic times; (b) as a generator of effective messages for the market, through brochures, web pages etc.; (c) as a Data mining tool for finding the most effective approaches for direct mail campaigns of existing customers or prospects (from purchased databases), and for call center activities etc.; (d) as a tool for one to one communications with typed prospects and prospects in general, either in person, or at the web page, or using targeted mails, or through call center. I.e. the full plethora of Mind Genomics uses is extremely wide, which opens many new horizons for this newly emerging technology. (This paragraph relates to two figures, this one and the next one.)

As the result of Mind Genomics Phase 1, a list of slogans was created that were appropriate for each of the customers' mind types. The figure accompanying this paragraph shows how Mind Genomics can, by creating correct list of "positive" and "negative" slogans (in this case for the "The personal touch" mind set), helped bank's employees to avoid negative messaging reducing them by 57 % and to employ suggested positive messaging increasing them by a staggering 460 %. As it might have been expected, such a correct "guessing" of the minds of the customers helped increase their satisfaction a great deal resulting in sizable increases of the bank's profits.

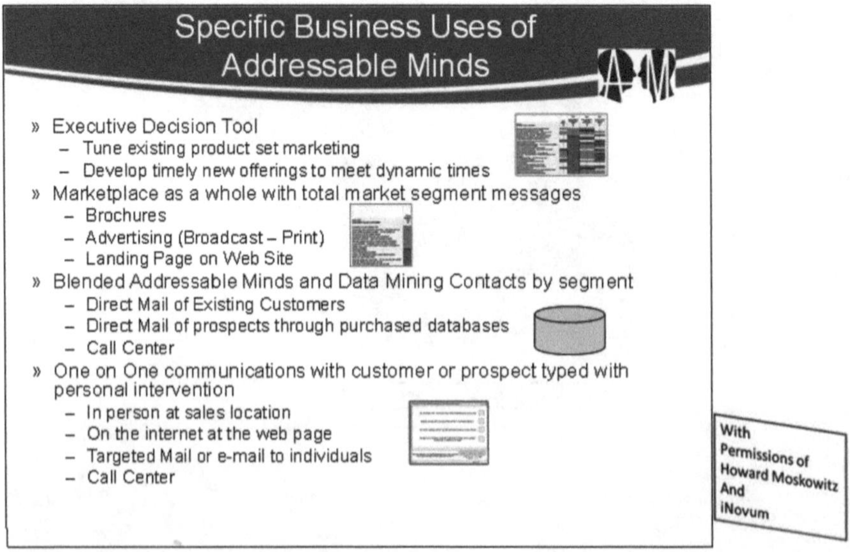
Specific Business Uses of Addressable Minds
» Executive Decision Tool
– Tune existing product set marketing
– Develop timely new offerings to meet dynamic times
» Marketplace as a whole with total market segment messages
– Brochures
– Advertising (Broadcast – Print)
– Landing Page on Web Site
» Blended Addressable Minds and Data Mining Contacts by segment
– Direct Mail of Existing Customers
– Direct Mail of prospects through purchased databases
– Call Center
» One on One communications with customer or prospect typed with personal intervention
– In person at sales location
– On the internet at the web page
– Targeted Mail or e-mail to individuals
– Call Center
With Permissions of Howard Moskowitz And iNovum

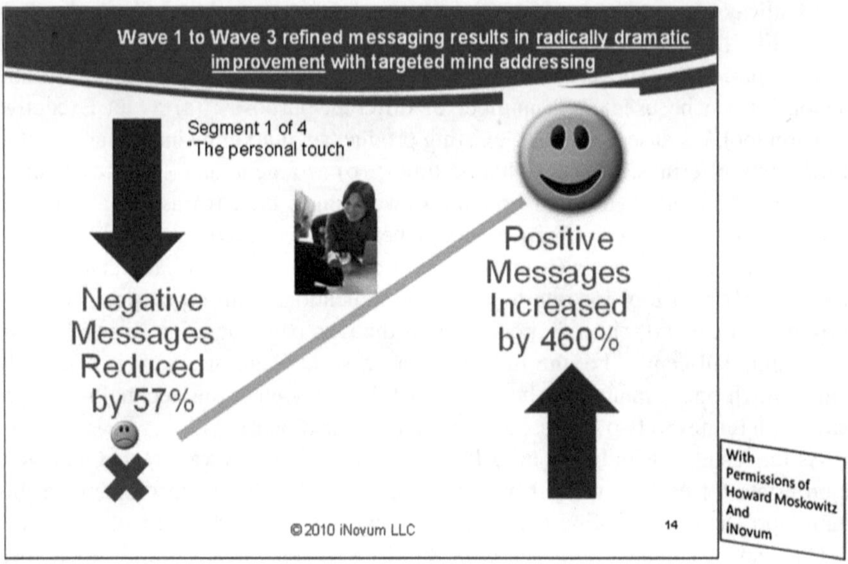
Wave 1 to Wave 3 refined messaging results in radically dramatic improvement with targeted mind addressing
Segment 1 of 4
"The personal touch"
Negative Messages Reduced by 57%
Positive Messages Increased by 460%
©2010 iNovum LLC
14
With Permissions of Howard Moskowitz And iNovum

Chapter 4
The Infrastructure Framework for Mind Genomics

Abstract This chapter links Mind Genomics with various techniques used to create electronic stores for electronic business on the Internet. Since Mind Genomics could exist only in the Internet-based environments, it is treated as an add-on on the top of some appropriate e-Business infrastructure. Notes were made in relation to three different pioneering techniques used for creation of electronic stores on the Internet, and the readers are pointed to more detailed sources, in a previous textbook of one of the authors of this textbook.

Keywords Mind genomics · Internet · Electronic stores · Electronic business on the Internet · e-Business infrastructure · Creation of e-Stores for small, medium, and large businesses

In most cases, Mind Genomics is used as a software layer on the top of some kind of infrastructure for sales over the Internet. Therefore, a question important for studying Mind Genomics is how to create a site for electronic sales over the Internet. Fundamentals of these issues are well documented by Milutinovic in "Infrastructure for Electronic Business on the Internet".

Basically, there are three major approaches: (a) one for extremely small businesses, including "one man bands", exemplified by systems like Yahoo!Store; (b) one for medium size businesses, pioneered by systems like ecBuilders (Milutinovic 2001); and (c) for extremely large businesses, exemplified by Microsoft Site Server Commerce Edition.

In the case of "small approaches", creation of an Internet store boils down to filling a form, which is very easy, and could be done by anyone. This is a quick and dirty approach. The good side is that it is easy to make it. The down side is that the capabilities are limited and no room is left for creativity of the business owner.

In the case of "medium approaches", creation of an Internet store boils down also to filling forms; however, in each step some room is left for the business owner to incorporate creative ideas. This typically means that linking capabilities are provided to other web sites, created by the owner or by the related businesses, worldwide.

V. Milutinovic and J. Salom, *Mind Genomics*, SpringerBriefs in Business,
DOI 10.1007/978-3-319-39733-7_4

In the case of "large approaches", creation of an Internet store boils down to making everything from scratch, which means that the business owner has lots of space for creativity, both in the domain of functionality and in the domain of presentation. Mind Genomics gives the best results if incorporated into the most complex approaches.

In all three cases above, five prerequisites have to be provided: (a) Internet merchant bank account, (b) web space, (c) secure payments certificate, (d) a contract with a provider of online transactions, and (e) electronic shopping cart. Getting an Internet merchant bank account is the responsibility of the business owner. For the other four prerequisites, the approach taken is determined by the selection of the infrastructure to be used. In the case of the "small approaches", everything is fixed and obtainable without the business being able to make selections. In the case of the "medium approaches", the business has an option, to take the defaults or to select its own preferences. In the case of the "large approaches", the business has full freedom in selecting the desired preferences.

For the background information of interest to Mind Genomics research, interested readers are referred to Moskowitz (2012), Gabay and Moskowitz (2015), Moskowitz et al. (2006), Zafeiropoulou (2015), Olmstead et al. (2015), Moss (2013), Columbia (2015), Cummings (2015), Carrell (2015), Oxford (2015), and Milutinovic (2001).

References

Carrell L (2015) Digital trends shaping the US marketing landscape. The Guardian. UK. Available at http://www.theguardian.com/media-network/2015/jun/01/trends-us-digital-marketing-landscape, [November 2015]

Columbia (2015) Statistical sampling and regression: simple linear regression. PreMBA analytical methods. Columbia University, USA. Available at http://ci.columbia.edu/ci/premba_test/c0331/s7/s7_6.html, [November 2015]

Cummings, K. (2015) Further reading | mindset segmentation. Available at http://www.kelliecummings.com/strategies/mindset-segmentation/further-reading-mindset-segmentation/, [November 2015]

Gabay G, Moskowitz H (2015) Mind genomics: what professional conduct enhances the emotional wellbeing of teens at the hospital? J Psychol Abnorm Child 4(3):2015

Milutinovic V (2001) Infrastructure for electronic business on the internet. Kluwer Academic Publishers, Boston, Dordreht, London

Moss M (2013) The extraordinary science of addictive junk food. The New York Times Magazine. USA. 20 Feb 2013

Moskowitz H, Gofman A, Beckley J, Ashman H (2006) Founding a new science: mind genomics. J Sens Stud 21:266–307

Moskowitz H (2012) Mind genomics: The experimental, inductive science of the ordinary, and its application to aspects of food and feeding. Physiol Behav 107(4):606–613

Olmstead M, Gilbert J et al (2015) In pursuit of the perfect peach: consumer-assisted selection of peach fruit traits. HortScience 50(8):1202–1212

Oxford (2015) Future trends and market opportunities in the world's largest 750 cities. Oxford economics. Oxford, United Kingdom, Available at https://www.oxfordeconomics.com/Media/Default/landing-pages/cities/oe-cities-summary.pdf, [November 2015]

Zafeiropoulou S, Nadan J (2015) Increasing the value of innovation consulting services in the technology age. Portland international conference on management of engineering and technology (PICMET), 2015 (pp 704–717)

Chapter 5
An Efficient Technological Support for Mind Genomics

Abstract This chapter brings attention to the fact that Mind Genomics creates a number of problems in the domain of BigData Analytics, and that these problems are effectively solvable only with modern DataFlow technology. One such DataFlow technology, coming from the Maxeler Technologies, is briefly overviewed and the interested readers are pointed to more detailed sources on the subject. This technology brings speedups of 20 to 200 or even higher in some cases, while the power consumption is about 20 times smaller; dimensions are also about 20 times smaller.

Keyword Mind genomics · DataFlow computing · Big data analytics in marketing · Maxeler technologies · Speedups · Power reductions

A major question in the implementation of on-line Mind Genomics is "what technological support to select for the on-line Mind Genomics, which is a data-intensive problem?" The answer is: "DataFlow SuperComputing for BigData."

The essence of DataFlow SuperComputing is that the code is written to configure the hardware, not to control the flow of data through the hardware. This fact defines its advantages and defines the related programming model.

DataFlow computers, compared to ControlFlow computers, offer speedups of 20–200 (even 2000 for some applications), power reductions of about 20, and size reductions of also about 20. However, as indicated, the programming paradigm is different, and has to be mastered.

The paradigm is best explained using Maxeler as an example (www.maxeler.com and appgallery.maxeler.com). Examples of dataflow success include not only BigData analytics typical of Mind Genomics, but also SignalProcessing, GeoPhysics, WeatherForecast, OilGas, DataEngineering, DataMining, etc. For example, a recent study from Tsinghua University in China reveals that, for Shallow Water Weather Forecast, which is a BigData problem, on the 1U level, the Maxeler DataFlow machine is 14 times faster than the Tianhe machine, which is rated #1 on the Top 500 list (based on Linpack, which is a small data benchmark).

V. Milutinovic and J. Salom, *Mind Genomics*, SpringerBriefs in Business,
DOI 10.1007/978-3-319-39733-7_5

The programming paradigm used for the Maxeler dataflow machines follows the principles of OpenSPL, established in 2014 by Stanford, Imperial of London, Tsinghua, and the University of Tokyo.

Further Readings

Trifunovic N, Milutinovic V et al (2016) The appgallery for big data supercomputing. J Big Data
Milutinovic V et al (2015) Guide to dataflow supercomputing. Springer
Milutinovic V et al (eds) (2015) Advances in computers: dataflow. Elsevier
Trifunovic N, Milutinovic V et al (2015) Paradigm shift in supercomputing: dataflow vs. controlflow. J Big Data
Jovanovic Z, Milutinovic V (2012) FPGA accelerator for floating-point matrix multiplication, The IET Comput Digit Tech 6(4):249–256
Flynn M, Mencer O, Milutinovic V et al (2013) Moving from petaflops to petadata. Commun ACM

Chapter 6
The Epilogue

Abstract This chapter includes concluding remarks and discusses possible epilogues that this newly emerging marketing technology could bring. References are given to cover relevant aspects of consequences in various domains. The major advantage of Mind Genomics is that it neither takes precious time from prospects (potential customers), nor it follows the prospects (in their ordinary activities on the Internet). It is based on a sophisticated math and on value exchange in critical moments of the marketing process. These characteristics make it attractive and promising.

Keyword Mind genomics · Emerging marketing technology · Consequences of mind genomics · Prospects · Internet · Value exchange

Innovation process is a step by step process. Once an important step up front is created, and a new innovation finds its way into the commercial world, it is difficult to imagine that another dramatical step/leap forward is possible. However, such steps keep happening. Mind Genomics is an example of one such step, unthinkable of until only a few years ago. What are the essential issues, unthinkable of, in Mind Genomics? Here is the answer: Modern targeted marketing methods collect information from users at the entry point into the system, and they cannot be extremely effective unless they track the behavior of the users. What was unthinkable of by the creators of these methods is that the effectiveness could be even higher without any entry point information collection and without any activities related to the tracking of the behavior of the customers. Yet, this is exactly what Mind Genomics is doing, due to a great innovation of Howard Moskowitz, a Harvard alumnus: Mind Genomics includes no information collection and no behavior tracking. It is based on sophisticated mathematics and statistics in its first phase, which refers to market analysis and on the principals of values exchange and fast Internet in its second phase, related to customer typing.

Innovation process rarely makes a lot of sense unless it results in an implementation which enables its sophistication to serve the human needs. A much more desirable scenario is the one in which the innovation creates a revolutionary leap

V. Milutinovic and J. Salom, *Mind Genomics*, SpringerBriefs in Business,
DOI 10.1007/978-3-319-39733-7_6

forward of the human wellbeing. Such achievements are possible only with the help of an appropriate enabler technology. That is exactly where the innovation like Mind Genomics needs an innovative technology like MultiScale DataFlow, exemplified in the recent research and developments of another great innovator, Oskar Mencer of Maxeler Technologies, a Stanford alumnus. Mind Genomics and MultiScale DataFlow could synergize in data analytics typical of giants like Google or Yahoo, Facebook or Tweeter, J.P. Morgan or The World Bank. It is absolutely impossible, at this time, to envision all the positive effects for the mankind that a synergy of the two (Mind Genomics and MultiScale DataFlow) could bring up. Needs of the users could be served much more effectively not only in business domains or other lucrative domains, but also in the domains of public health, public happiness, public and individual quality of life, public and individual understanding of the environment around, etc …

Innovation is a process that requires sacrifices, and its success is typically paved by blood, sweat, and tears. In a number of schools of thought, in a number of different cultures, at various times, deep thinkers had created a wisdom which is very applicable both to Mind Genomics and MultiScale DataFlow, and especially to the synergy of the two. This wisdom, in its various incarnations, reads as following: Each great innovation passes through the four inevitable phases. In the first phase, the innovation is ridiculed at. In the second phase, the opponents, sensing that they are slowly losing their grounds, try to kill the innovation, in their conscience or sub-conscience despair, related to the facts that either their lives were tuned to something obsolete, or their business is becoming obsolete. In the third phase, after an undisputable success is reached, many try to arrogate it; often times, that is done very aggressively, and typically by those who were the most aggressive critics in the previous phase. Finally, in the fourth phase, which often times comes after the actors of all the previous phases are already dead, the new generation believes that the innovation has existed forever, maybe even all the way back since the times of the ancient Greeks or Romans. Consequently, developers of great innovations have constantly to think about catalytic mechanisms that would speed up the innovation's flow through initial, painful phases.

We hope this book is one such catalytic mechanism, a small one, with elements of a symbiotic stress on the synergy between two innovations that could serve to each other as accelerators through the above described process. In order to make this book readable easily, after the initial engineering-style explanations using conventional narrator methodologies, the business oriented case studies were explained using a recently more and more used, PowerPoint-like narrator style. Often, once the readers understood the essence, they would have no patience to read carefully all the tedious details of the follow-up case studies. With the PowerPoint-like method used here, the readers should be able to comprehend quickly the essence of all the presented case studies.

6.1 Further Readings About Dataflow Supercomputing, Part #1: Educational Books and Papers

Books:

Milutinovic V, Salom J, Trifunovic N, Giorgi R (2015) Guide to dataflow supercomputing. Springer

Milutinovic V (ed) (2015) Advances in computers: dataflow. Elsevier

Papers:

Flynn M, Mencer O, Milutinovic V et al (2013) Moving from petaflops to petadata. Commun ACM

Milutinovic V et al (2015) Paradigm shift in supercomputing: dataflow vs. controlflow. J Big Data

6.2 Further Readings About Dataflow Supercomputing, Part #2: Research Papers and Books

Papers:

Milutinovic D, Milutinovic V (1996) Mapping of interconnection networks for parallel processing onto the sea-of-gates VLSI. IEEE Computer 29(6)

Jovanovic Z, Milutinovic V (2012) FPGA accelerator for floating-point matrix multiplication. The IET Comput Digit Tech 6(4):249–256

Flynn M et al (2013) Moving from petaflops to petadata. Commun ACM 56(5):39–43

Trifunovic N, Milutinovic V, Salom J, Kos A (2015) Paradigm shift in big data supercomputing: dataflow vs. controlflow. J Big Data 2(4)

Books:

Milutinovic V (ed) (1988) Computer architecture (Chap 9, Dataflow computation, Dennis J), North Holland

Milutinovic V (ed) (1989) High-level language computer architecture (Chap 9, Dataflow machines, Gaudiot JL), Computer Science Press

FSC
www.fsc.org
MIX
Papier | Fördert
gute Waldnutzung
FSC® C083411

Kolibri 360 GmbH
Ferdinand-Jühlke-Straße 7
99095 Erfurt, Deutschland
produktsicherheit@kolibri360.de